Leonid Berlyand and Pierre-Emmanuel Jabin
Mathematics of Deep Learning

Also of Interest

Deep Learning for Cognitive Computing Systems
Technological Advancements and Applications
Edited by M. G. Sumithra, Rajesh Kumar Dhanaraj, Celestine Iwendi, Anto Merline Manoharan, 2022
ISBN 978-3-11-075050-8, e-ISBN (PDF) 978-3-11-075058-4
Smart Computing Applications
Edited by Prasenjit Chatterjee, Dilbagh Panchal, Dragan Pamucar, Sharfaraz Ashemkhani Zolfani
ISSN 2700-6239, e-ISSN 2700-6247

Deep Learning
Research and Applications
Edited by Siddhartha Bhattacharyya, Vaclav Snasel, Aboul Ella Hassanien Satadal Saha, B. K. Tripathy, 2020
ISBN 978-3-11-067079-0, e-ISBN (PDF) 978-3-11-067090-5
De Gruyter Frontiers in Computational Intelligence
Edited by Siddhartha Bhattacharyya
ISSN 2512-8868, e-ISSN 2512-8876

Algorithms
Design and Analysis
Sushil C. Dimri, Preeti Malik, Mangey Ram, 2021
ISBN 978-3-11-069341-6, e-ISBN (PDF) 978-3-11-069360-7

Mathematical Logic
An Introduction
Daniel Cunningham, 2023
ISBN 978-3-11-078201-1, e-ISBN (PDF) 978-3-11-078207-3

Advanced Mathematics
An Invitation in Preparation for Graduate School
Patrick Guidotti, 2022
ISBN 978-3-11-078085-7, e-ISBN (PDF) 978-3-11-078092-5

Leonid Berlyand and Pierre-Emmanuel Jabin

Mathematics of Deep Learning

An Introduction

DE GRUYTER

Mathematics Subject Classification 2010
47-01, 47A10, 47A30, 47A35, 47A56, 47A60, 47B07, 47B25, 46-01, 46A30, 46A35, 46A45, 46E15

Authors
Prof. Leonid Berlyand
Penn State University
Department of Mathematics
University Park PA 16802
USA
lvb2@psu.edu

Prof. Pierre-Emmanuel Jabin
Penn State University
Department of Mathematics
109 McAllister
University Park PA 16802
USA
pejabin@psu.edu

ISBN 978-3-11-102431-8
e-ISBN (PDF) 978-3-11-102555-1
e-ISBN (EPUB) 978-3-11-102580-3

Library of Congress Control Number: 2023931034

Bibliographic information published by the Deutsche Nationalbibliothek
The Deutsche Nationalbibliothek lists this publication in the Deutsche Nationalbibliografie; detailed bibliographic data are available on the Internet at http://dnb.dnb.de.

Cover image: metamorworks / iStock / Getty Images Plus
Typesetting: VTeX UAB, Lithuania
Printing and binding: CPI books GmbH, Leck

www.degruyter.com

Contents

1 About this book

This book aims at providing a mathematical perspective on some key elements of the so-called deep neural networks (DNNs). Much of the interest in deep learning has focused on the implementation of DNN-based algorithms. Our hope is that this compact textbook will offer a complementary point of view that emphasizes the underlying mathematical ideas. We believe that a more foundational perspective will help to answer important questions that have only received empirical answers so far.

Our goal is to introduce basic concepts from deep learning, e. g., mathematical definitions of DNNs, loss functions, the backpropagation algorithm, etc., in a rigorous mathematical fashion. We attempt to identify for each concept the simplest setting that minimizes technicalities but still contains the key mathematics.

The book focuses on deep learning techniques and introduces them almost immediately. Other techniques such as regression and support vector machines are briefly introduced and used as a stepping stone to explain basic ideas of deep learning. Throughout these notes, the rigorous definitions and statements are supplemented by heuristic explanations and figures.

The book is organized so that each chapter introduces a key concept. When teaching this course, some chapters could be presented as a part of a single lecture, whereas others contain more material and would take several lectures.

This book is based on two special topic courses on the mathematics of deep learning taught at Penn State, first for PhD students and two years later also for senior undergraduates and master students. The idea behind those courses was to translate some of the computer science concepts for deep learning into some simple mathematical language that would be more accessible to math students. The lecture notes from these courses became the ground material for this book.

The preparation of this manuscript has been a fun undertaking, involving numerous Penn State students. Here is the list of PhD students to whom we are extremely grateful for many insightful discussions in the preparation of the manuscript: Hai Chi, Robert Creese, Spencer Dang, Jonathan Jenkins, Oleksii Krupchytskyi, Alex Safsten, Yitzchak Shmalo, and Datong Zhou. Alex, Datong, Hai, Robert, Spencer, and Yitzchak took the courses in deep learning and helped in typing the lecture notes. Furthermore, Alex, Datong, and Spencer provided invaluable help in creating figures. It is also a great pleasure to thank Alex and Yitzchak for assisting in the preparation of the exercises for this book. Alex, Jonathan, Oleksii, Spencer, and Yitzchak provided very efficient and timely help in proofreading the manuscript.

We express our special gratitude to Alex Safsten for his contributions to both our teaching and research on deep learning and, in particular, for suggesting several examples for these lecture notes.

Finally, we acknowledge partial support of our work on the mathematics of deep learning from the National Science Foundation.

https://doi.org/10.1515/9783111025551-001

2 Introduction to machine learning: what and why?

2.1 Some motivation

Machine learning-based artificial intelligence has entered our daily lives in many various ways, such as:
- self-driving cars,
- speech and facial recognition,
- Google search, and
- product (Amazon) and content (YouTube) recommendations,

with ever further reaching impacts.

What may be less apparent is the increasing role taken by machine learning algorithms in classical scientific computing questions. A recent seminal example concerned the highly accurate protein structure prediction with AlphaFold in [22]. Proteins are long sequences of amino acids – a type of organic molecule (organic molecules are based on carbon–hydrogen bonds). Proteins are critical to many biochemical processes in living organisms, with a prominent example being DNA replication for cell division.

The 3D structure of a protein is necessary for the biochemical functions of the protein in the organism. However, a long-standing problem (since 1960) is the question of how proteins fold, that is, how does the amino acid sequence composing the protein determine its 3D structure? Knowledge of a protein's structure is essential to understand its properties.

Mathematically, the protein folding problem can be formulated either as a dynamical problem or as an energy minimization problem: the protein assumes the shape that minimizes the energy stored in atomic bonds. This problem presents a great computational challenge because proteins are extremely complicated molecules consisting of tens or hundreds of thousands of atoms. Moreover, the problem of cataloging how all proteins fold through laboratory experiments is infeasible because of the many ways a protein can fold and the huge number of proteins. In 2021, Google's machine learning algorithm *AlphaFold* was the first computational method to achieve an accuracy of more than 95 % in predicting protein folding.

2.2 What is machine learning?

We use several common terms such as artificial intelligence and machine learning. Because there is no consensus on the exact meaning of those terms, we specify below in which sense they are used in this book.
- *Artificial intelligence* is (according to Wikipedia) "intelligence demonstrated by machines as opposed to natural intelligence demonstrated by animals including hu-

https://doi.org/10.1515/9783111025551-002

mans." One of the ideas behind artificial intelligence is to try to mimic the cognitive function of a brain, e. g., "learning" and problem solving.
- *Machine learning* is a subfield of artificial intelligence. To explain the name, we note that in a non-learning algorithm, all parameters and methods used to produce the output are predetermined by the programmer. In contrast, machine learning algorithms use at least some parameters that are not fixed. Instead, the algorithms improve themselves ("learn") by refining their parameters and methods through experience such as comparison of its output with known data. Because of that, a machine learning algorithm typically goes through a distinct training phase before it is actually used to solve problems.

3 Classification problem

As a first in-depth example of an important practical problem, consider *the classification problem* of placing objects into categories, also called classes. For example, one may classify images of handwritten digits by determining which digit is depicted.

In a rigorous mathematical formulation of this problem, objects are represented as points $\boldsymbol{s} \in \mathbb{R}^n$. For example, images of handwritten digits can be represented by a vector of their n pixel values, and the ith pixel is represented by the coordinate $\boldsymbol{s}_i$. In a simple setting of grayscale (black and white) images, $\boldsymbol{s}_i$, $i = 1, \ldots, n$, takes a value in $[0, 1]$ which represents the lightness of the pixel from 0 (black) to 1 (white). We denote by $S \subset \mathbb{R}^n$ the set of all objects to be considered; S may be finite or infinite. In practice, one starts from some finite subset T of the entire set of objects S. Objects in T are already classified. Then this information is used to classify objects from S, which is typically much larger than T. The set T is called the *training set* because the classification method is "trained" on this set to classify objects which are not in T.

The classes correspond to subsets $S_j \subset S$ indexed by integers $j = 1, \ldots, m$. For example, in the above problem of handwritten digits, there are clearly ten disjoint classes. They are naturally pairwise disjoint,

$$S_j \cap S_k = \emptyset \quad \text{for } j \neq k, \tag{3.1}$$

and together they make up all of S:

$$\bigcup_{j=1}^{m} S_j = S. \tag{3.2}$$

A function $\hat{\phi} : S \to \{1, \ldots, m\}$ which correctly maps each object to the index of its class is called an *exact classifier* (if the classifier is not exact, objects may be incorrectly mapped). A function which assigns (maps) each object to a class is called a *hard classifier*. In other words, the range of the hard classifier function is a collection of classes.

In contrast, it is frequently more convenient to use a *soft classifier*. Soft classifiers assign to each object $\boldsymbol{s} \in S$ a vector

$$p(s) = (p_1(s), \ldots, p_m(s)), \tag{3.3}$$

where each component $p_i(s)$ represents the predicted probability that $\boldsymbol{s}$ is in class i. An exact soft classifier would map each object $\boldsymbol{s}$ in class i to the vector $(p_1(s), \ldots, p_m(s))$, where $p_{i(s)} = 1$ and $p_j(s) = 0$ for $j \neq i(s)$, where

$$i(s) = \hat{\phi}(s) \tag{3.4}$$

is the index of the correct class (which is known for $s \in T$) of object s.

https://doi.org/10.1515/9783111025551-003

In practical applications, it is typically not possible to find the exact classifier. Instead, the goal is to construct an approximation which depends on a set of parameters $\boldsymbol{\alpha}$ and optimize the choice of these parameters. This can of course lead to a hard or a soft classifier. In that regard, it is useful to note that any *soft approximate classifier* $\phi : \mathbb{R}^n \to \mathbb{R}^m$ can be transformed into a hard classifier $\bar{\phi} : \mathbb{R}^n \to \{1, \dots, m\}$ by taking $\bar{\phi}(s) = i$ if $p_i(s)$ is the largest component of $\phi(s)$.

4 The fundamentals of artificial neural networks

4.1 Basic definitions

In this book, we focus on *deep learning*, which is a type of machine learning based on *artificial neural networks* (ANNs), which involve several layers of so-called neuron functions. This kind of architecture was loosely inspired by the biological neural networks in animals' brains.

The key building block of ANNs is what we call here a *neuron function*. A neuron function is a very simplified mathematical representation of a biological neuron dating back to [34]. This naturally leads to various ways of organizing those neuron functions into interacting networks that are capable of complex tasks.

The perceptron, introduced in [43] (see also [35]), is one of the earliest one layer ANNs. Nowadays, multiple layers are typically used instead because neurons in the same layer are not connected, but neurons can interact with neurons in other layers.

We present in this chapter a rather common architecture for classifiers, which is based on so-called *feedforward networks*, which are networks with no cycles. A "cycle" means that the output of one layer becomes the input of a previous layer. More complex networks can also be used with some cycles between layers (so-called *recurrent* neural networks; see for example [11, 17]).

Finally as one readily sees, ANNs typically require a very large number of parameters. This makes identifying the best choice for those coefficients delicate and leads to the important question of learning the parameters, which is discussed later in this book.

Definition 4.1.1. A *neuron function* $f : \mathbb{R}^n \to \mathbb{R}$ is a mapping of the form (see Fig. 4.1)

$$f(\boldsymbol{x}) = \lambda(\boldsymbol{\alpha} \cdot \boldsymbol{x} + \beta), \tag{4.1}$$

where $\lambda : \mathbb{R} \to \mathbb{R}$ is a continuous non-linear function called the *activation function*, $\boldsymbol{\alpha} \in \mathbb{R}^n$ is a vector of *weights*, and scalar $\beta \in \mathbb{R}$ is called the *bias*. Here, $\boldsymbol{\alpha} \cdot \boldsymbol{x}$ is the inner product on $\mathbb{R}^n$.

A typical example of an activation function is *ReLU* (*Re*ctified *L*inear *U*nit), defined as follows (see Fig. 4.2):

$$\mathrm{ReLU}(x) = \begin{cases} x & x > 0, \\ 0 & x \leq 0. \end{cases} \tag{4.2}$$

ReLU is a very simple non-linear function composed of two linear functions that model the threshold between an "inactive" and an "active" state of the neuron.

ReLU is a simple way to model a neuron in a brain which has two states: resting (resting potential) and active (firing). Incoming impulses from other neurons can change the state from resting to active, but the impulse needs to reach a certain threshold first.

https://doi.org/10.1515/9783111025551-004

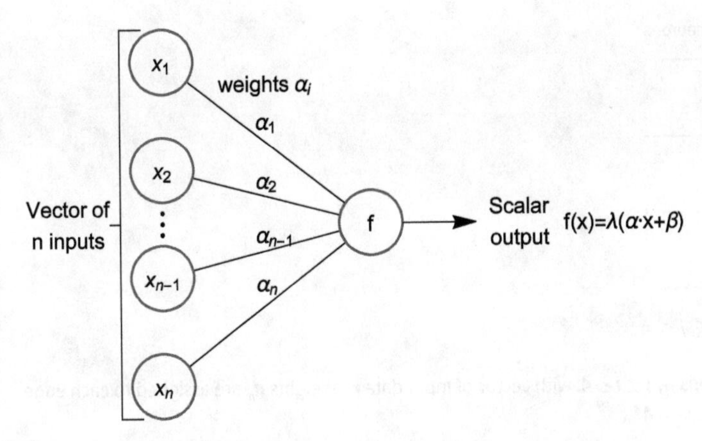

Figure 4.1: A neuron function input $\boldsymbol{x} \in \mathbb{R}^n$, vector of weights $\boldsymbol{\alpha} = (\alpha_1, \ldots, \alpha_n)$, bias β, and output scalar $f(x) = \lambda(\boldsymbol{\alpha} \cdot \boldsymbol{x} + \beta)$.

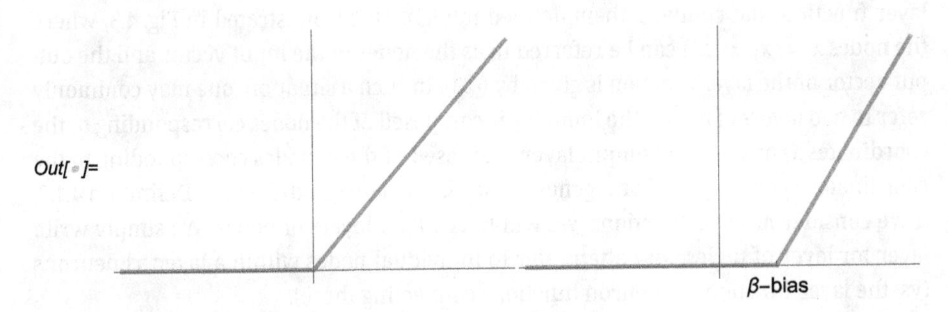

Figure 4.2: ReLU function. β, "bias" constant.

Thus, ReLU's change from constant 0 to a linear function at $x_{\text{thresh}} = \beta$ reflects the change from resting to active at this threshold.

Since the output of a neuron function is a single number, we can combine several neurons to create a vector-valued function called a *layer function.*

Definition 4.1.2. A *layer function* $g : \mathbb{R}^n \to \mathbb{R}^m$ is a mapping of the form

$$g(\boldsymbol{x}) = (f_1(\boldsymbol{x}), f_2(\boldsymbol{x}), \ldots, f_m(\boldsymbol{x})), \tag{4.3}$$

where each $f_i : \mathbb{R}^n \to \mathbb{R}$ is a neuron function of the form (4.1) with its own vector of parameters $\boldsymbol{\alpha}_i = (\alpha_{i1}, \ldots, \alpha_{in})$ and biases β_i, $i = 1, \ldots, m$.

Remark 4.1.1. When $m = 1$ in (4.3), the layer function reduces to a neuron function.

Remark 4.1.2. When discussing layers, it is important to distinguish between the layer nodes and the layer function that connects them. There are two columns of nodes in Fig. 4.3, which are commonly referred to as two layers. However, according to Definition 4.1.2, Fig. 4.3 depicts a single layer function. Thus, it is important to distinguish be-

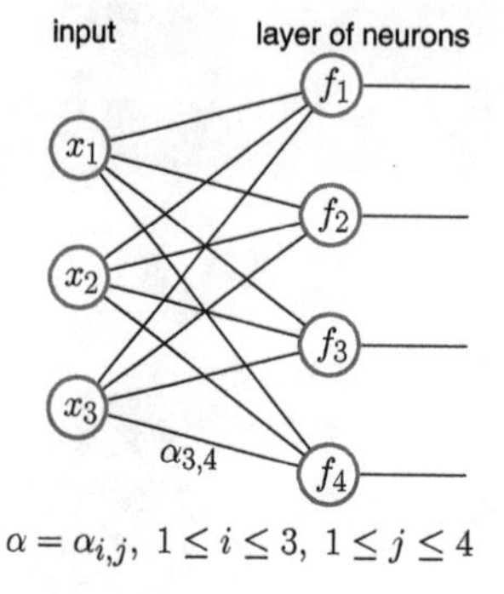

Figure 4.3: Layer of neurons f_I, $1 \le i \le 4$, with vector of input data x_j. Weights a_{ij} are assigned to each edge connecting x_j and f_i, $n = 3$, $m = 4$.

tween columns of nodes in diagrammatic representations of layers as in Fig. 4.3 and the layer function that connects them defined in (4.3). This is illustrated in Fig. 4.3, where the nodes $\mathbf{x} = (x_1, x_2, x_3)$ can be referred to as the nodes of the input vector and the output vector of the layer function is given by (4.3). In such a situation, one may commonly refer to two layers of nodes: the input layer composed of the nodes corresponding to the coordinates x_i of x and the output layer composed of the m nodes corresponding to the coordinates y_i of $y = g(x)$. For a general multilayer network defined in Definition 4.1.3, if we consider M layer functions, we will have $M + 1$ layers of nodes. We simply write layer for layer of nodes, and often refer to individual nodes within a layer as neurons (vs. the layer functions or neuron functions connecting those).

Thus, the layer function is determined by a *matrix of parameters,*

$$A = \begin{pmatrix} \alpha_{11} & \cdots & \alpha_{1n} \\ \vdots & \ddots & \vdots \\ \alpha_{m1} & \cdots & \alpha_{mn} \end{pmatrix}, \tag{4.4}$$

and a *vector of biases,*

$$\boldsymbol{\beta} = \begin{pmatrix} \beta_1 \\ \vdots \\ \beta_m \end{pmatrix}. \tag{4.5}$$

Hence, (4.3) may be written

$$g(\mathbf{x}) = \bar{\lambda}(A\mathbf{x} + \boldsymbol{\beta}), \tag{4.6}$$

where $\bar{\lambda} : \mathbb{R}^m \to \mathbb{R}^m$ is the *vectorial activation function* defined as

$$\bar{\lambda}(x_1, \ldots, x_m) = (\lambda(x_1), \ldots, \lambda(x_m)) \tag{4.7}$$

for a scalar activation function λ as in Definition 4.1.1.

See Fig. 4.3 for a diagram of a layer function. This figure shows a layer as a graph with two columns of nodes. The right column depicts neuron functions in the layer, while the left column depicts three real numbers (data) which are input to the layer. The edge connecting the ith input with the jth neuron is multiplied by the parameter value a_{ij} from the matrix of parameters.

Definition 4.1.3. An *artificial neural network* (ANN) is a function $h : \mathbb{R}^n \to \mathbb{R}^m$ of the form

$$h(\mathbf{x}) = h_M \circ h_{M-1} \circ \cdots \circ h_1(\mathbf{x}), \quad M \geq 1, \tag{4.8}$$

where each $h_i : \mathbb{R}^{n_{i-1}} \to \mathbb{R}^{n_i}$ is a layer function (see Definition 4.1.2) with its own matrix of parameters A^i and its own vector of biases $\boldsymbol{\beta}^i$.

Fig. 4.4 shows an ANN composed of two layers. The layer function in Fig. 4.4 between input and output is called a *hidden layer* (hidden from the user) because its output is passed to another layer, not directly to the user. The number of neurons n_i in the ith layer is called the layer's *width*, while the total number M of layers in an ANN is the *depth* of the ANN. The numbers $n_1, \ldots, n_M$ and M comprise the *architecture* of this network. ANNs with more than one layer are referred to as *deep neural networks* (DNNs).

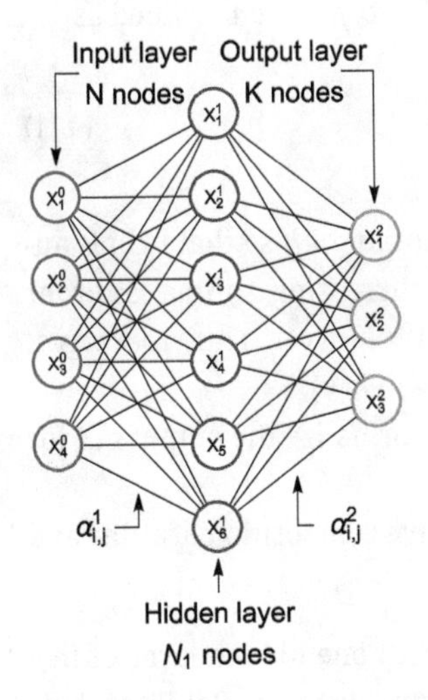

Figure 4.4: Simple network with input layer and two layer functions, $n = n_0 = 4$, $n_1 = 6$, $m = n_2 = 3$.

This network is called *fully connected* because for all but the last layer, each neuron provides input to each neuron in the next layer. That is, each node in each column of the graph in Fig. 4.4 is connected to each node in the next column. Finally, we mention that the last layer, corresponding to the result of the last layer function h_M, is commonly called the *output layer*. The input vector $\mathbf{x}$ is commonly referred to as the *input layer*.

Neither the input layer nor the output layer is a layer function in the sense of Definition 4.1.2, and both are instead layers of nodes (cf. Remark 4.1.2).

4.2 ANN classifiers and the softmax function

An ANN can be used to build a soft classifier function. Recall that a soft classifier outputs a list of probabilities that the input belongs to each class. That is, if $\phi(\boldsymbol{x}) = (\phi_1(\boldsymbol{x}), \ldots, \phi_m(\boldsymbol{x}))$ is a soft classifier, we should have for every $\boldsymbol{x}$

$$0 \le \phi_i(\boldsymbol{x}) \le 1,\ i = 1, \ldots m, \quad \text{and} \quad \sum_{i=1}^{m} \phi_i(\boldsymbol{x}) = 1. \tag{4.9}$$

For ANNs of the form given in Definition 4.1.3, (4.9) is difficult to satisfy. Therefore, we define an *ANN classifier* as follows.

Definition 4.2.1. An *ANN classifier* is a function $\tilde{h} : \mathbb{R}^n \to \mathbb{R}^m$ of the form

$$\tilde{h}(\mathbf{x}) = \sigma \circ h(\mathbf{x}), \tag{4.10}$$

where $h : \mathbb{R}^n \to \mathbb{R}^m$ is an ANN and $\sigma : \mathbb{R}^m \to \mathbb{R}^m$ is the *softmax function*, defined as

$$\sigma(\mathbf{y}) = \left(\frac{e^{y_1}}{\sum_{i=1}^{m} e^{y_i}}, \ldots, \frac{e^{y_m}}{\sum_{i=1}^{m} e^{y_i}} \right). \tag{4.11}$$

Observe that the softmax function "normalizes" the output of $\tilde{h}$ so that (4.9) is automatically satisfied. Moreover, since $y \mapsto e^y$ is monotone increasing, softmax is order preserving in the sense that if $\boldsymbol{y} \in \mathbb{R}^m$ and $y_i > y_j$, then $(\sigma(\boldsymbol{y}))_i > (\sigma(\boldsymbol{y}))_j$. Thus, the largest component of $h(\boldsymbol{x})$ is also the class with highest probability in $\tilde{h}(\boldsymbol{x})$.

The softmax function is of course not the only way to normalize the output but for classification problems it is very commonly used.

In summary, an ANN is a function obtained by an iterative composition of affine and non-linear functions.

Example 4.2.1. Consider an ANN function $h : \mathbb{R}^3 \to \mathbb{R}^2$ with one hidden layer of five neurons. Suppose that all neuron functions in both function layers use ReLU for their activation functions and have the following matrices of parameters and bias vectors:

$$A^1 = \begin{pmatrix} 4 & 2 & -5 \\ 5 & 4 & 6 \\ 0 & 3 & 2 \\ 2 & 2 & -3 \\ -7 & 1 & -4 \end{pmatrix}, \quad A^2 = \begin{pmatrix} -2 & 6 & 2 & -5 & 0 \\ -3 & 0 & 6 & 1 & 2 \end{pmatrix}, \tag{4.12}$$

$$\boldsymbol{b}^1 = \begin{pmatrix} -2 \\ 3 \\ 5 \\ -2 \\ 7 \end{pmatrix}, \quad \boldsymbol{b}^2 = \begin{pmatrix} -3 \\ -1 \end{pmatrix}. \tag{4.13}$$

Calculate the value of the ANN function h at the point $(1, 0, 6)$.

Solution: We may write $h = g_2 \circ g_1$, where g_1 is the first layer and g_2 is the second layer. The first layer is

$$g_1(\mathbf{x}) = \bar{\lambda}(A^1\mathbf{x} + \boldsymbol{b}^1), \tag{4.14}$$

where $\bar{\lambda}$ is the vectorial ReLU activation function. Therefore,

$$g_1(1,0,6) = \bar{\lambda}\left[\begin{pmatrix} 4 & 2 & -5 \\ 5 & 4 & 6 \\ 0 & 3 & 2 \\ 2 & 2 & -3 \\ -7 & 1 & -4 \end{pmatrix}\begin{pmatrix} 1 \\ 0 \\ 6 \end{pmatrix} + \begin{pmatrix} -2 \\ 3 \\ 5 \\ -2 \\ 7 \end{pmatrix}\right] \tag{4.15}$$

$$= \bar{\lambda}\left[\begin{pmatrix} -26 \\ 41 \\ 12 \\ -16 \\ -31 \end{pmatrix} + \begin{pmatrix} -2 \\ 3 \\ 5 \\ -2 \\ 7 \end{pmatrix}\right] \tag{4.16}$$

$$= \bar{\lambda}\left[\begin{pmatrix} -28 \\ 44 \\ 17 \\ -18 \\ -24 \end{pmatrix}\right] \tag{4.17}$$

$$= \begin{pmatrix} 0 \\ 44 \\ 17 \\ 0 \\ 0 \end{pmatrix}. \tag{4.18}$$

The second layer is

$$g_2(\boldsymbol{x}) = \bar{\lambda}(A^2\boldsymbol{x} + \boldsymbol{b}^2). \tag{4.19}$$

Therefore,

$$h(1,0,6) = g_2 \circ g_1(1,0,6) \tag{4.20}$$

$$= g_2(0, 44, 17, 0, 0) \tag{4.21}$$

$$= \bar{\lambda}\left[\begin{pmatrix} -2 & 6 & 2 & -5 & 0 \\ -3 & 0 & 6 & 1 & 2 \end{pmatrix}\begin{pmatrix} 0 \\ 44 \\ 17 \\ 0 \\ 0 \end{pmatrix} + \begin{pmatrix} -3 \\ -1 \end{pmatrix}\right] \tag{4.22}$$

$$= \bar{\lambda}\left[\begin{pmatrix} 298 \\ 102 \end{pmatrix} + \begin{pmatrix} -3 \\ -1 \end{pmatrix}\right] \tag{4.23}$$

$$= \bar{\lambda}\left[\begin{pmatrix} 295 \\ 101 \end{pmatrix}\right] \tag{4.24}$$

$$= \begin{pmatrix} 295 \\ 101 \end{pmatrix}. \tag{4.25}$$

4.3 The universal approximation theorem

One of the reasons for the success of deep learning methods stems from the fact that ANN functions can approximate a wide class of continuous functions. To illustrate this, we first present the approximation theorem from [19].

Theorem 4.3.1 (Universal approximation theorem). *Let $K \subset \mathbb{R}^n$ be a closed and bounded set and let $f : K \to \mathbb{R}^m$ be a continuous function such that its components are all non-negative: $f_i(\boldsymbol{x}) \geq 0$ for all $\boldsymbol{x} \in K$ and $1 \leq i \leq m$. Then for any $\varepsilon > 0$ there exists an ANN $h : \mathbb{R}^n \to \mathbb{R}^m$ with two function layers (i. e., three layers with one hidden layer) using the ReLU activation function such that*

$$\sup_{x \in K} \|h(\boldsymbol{x}) - f(\boldsymbol{x})\| < \varepsilon. \tag{4.26}$$

There are several versions of the universal approximation theorem; we refer again to [19] for a more general version.

While the universal approximation theorem guarantees that a positive continuous function may be approximated by an ANN with one hidden layer, it does not make any mention of how large that hidden layer must be. In practice, a good approximation may indeed require this hidden layer to be quite big, too large even for powerful computers to handle. This is where having a deeper network is useful, as many hidden layers, even if they are not so wide, may be able to perform the same approximation with a far lower total number of neuron functions.

Another version of the universal approximation theorem [38] applies to functions of the form $L_2 \circ \bar{\lambda} \circ L_1$, where L_1 and L_2 are affine maps (of the form $A\boldsymbol{x} + \boldsymbol{b}$) and $\bar{\lambda} : \mathbb{R}^k \to \mathbb{R}^k$ is the componentwise application of a non-linear activation function λ (see Definition 4.1.1):

$$\bar{\lambda}(x_1,\ldots,x_k) = (\lambda(x_1),\ldots,\lambda(x_k)). \tag{4.27}$$

Note that $\bar{\lambda}\circ L_1$ is a layer function, whereas $L_2\circ\bar{\lambda}\circ L_1$ is not an ANN function (4.8) according to our definitions because L_2 is not composed on the left with an activation function (an ANN would be given by $\bar{\lambda}\circ L_2\circ\bar{\lambda}\circ L_1$). Thus, only one layer function with its activation function composed with one affine map is needed for good approximations, but two layer functions also do the job.

Theorem 4.3.2 (Universal approximation theorem). *Let $K \subset \mathbb{R}^n$ be a closed and bounded set and let $\lambda : \mathbb{R} \to \mathbb{R}$ be a non-linear function which is not a polynomial. Let $f : K \to \mathbb{R}^m$ be a continuous function. For any $\varepsilon > 0$ there exist $k \in \mathbb{N}$ and affine functions $L_1 : \mathbb{R}^n \to \mathbb{R}^k$ and $L_2 : \mathbb{R}^k \to \mathbb{R}^m$ such that*

$$\sup_{x\in K}\|f(\mathbf{x}) - L_2\circ\bar{\lambda}\circ L_1(\mathbf{x})\| < \varepsilon, \tag{4.28}$$

where $\bar{\lambda}$ is the componentwise action of λ as in (4.27).

Question: Why can λ not be a polynomial?

Answer: For simplicity, we assume that the dimensions of the input and output are 1, i. e., $n = m = 1$ in Theorem 4.3.2. Suppose that λ is a polynomial of degree $d \in \mathbb{N}$. Then L_1 is an affine map $\mathbb{R} \to \mathbb{R}^k$ (k is the width of the hidden layer) and L_2 is an affine map $\mathbb{R}^k \to \mathbb{R}$, so $L_1(x) = \mathbf{a}_1 x + \mathbf{b}_1$, where $\mathbf{a}_1, \mathbf{b}_1 \in \mathbb{R}^k$, and $L_2(\mathbf{y}) = \mathbf{a}_2\cdot\mathbf{y} + b_2$, where $\mathbf{a}_2 \in \mathbb{R}^k$ and $b_2 \in \mathbb{R}$. Thus $L_2\circ\bar{\lambda}\circ L_1$ is a linear combination of neuron functions of the form $\bar{\lambda}(\mathbf{a}_1 x + \mathbf{b}_1)$. Clearly, a polynomial of degree d composed with an affine function is still a polynomial of degree d. Furthermore, a linear combination of polynomials of degree d is a polynomial of degree d or less. Therefore, $L_2\circ\bar{\lambda}\circ L_1$ is a polynomial of degree $\le d$ and can at best approximate any function which can be approximated by d-degree polynomials. Recall that the Weierstrass approximation theorem [39] states that arbitrarily close approximations can be achieved by polynomial functions, and in general, an arbitrarily large degree may be needed.

We show below that the function $f(x) = \cos(x)$ in the interval $K = [0, 4\pi]$ cannot be approximated by polynomials with degree ≤ 3. This function has five points where it reaches a local maximum of 1 or minimum of -1: $s_0 = 0$, $s_1 = \pi$, $s_2 = 2\pi$, $s_3 = 3\pi$, and $s_4 = 4\pi$. Note that with this notation, if i is even, then $f(s_i) = 1$, and if i is odd, then $f(s_i) = -1$.

Suppose that $\cos(x)$ can be approximated by a polynomial of degree 3 in the interval K. Then for any $\varepsilon > 0$, there exists a cubic polynomial $p(x)$ such that

$$|\cos(x) - p(x)| < \varepsilon \quad \forall x \in K. \tag{4.29}$$

Let $\varepsilon = 1/2$ and let $p(x)$ be a cubic polynomial such that (4.29) holds. Then $p(x)$ has at most three roots in K. We will consider the case that p has exactly three roots in K (with the case of fewer roots following by the same argument). Let us label the roots of p as $r_1 < r_2 < r_3$. We may decompose the interval K into four subintervals:

$$K = K_1 \cup K_2 \cup K_3 \cup K_4 := [-4\pi, r_1) \cup [r_1, r_2) \cup [r_2, r_3) \cup [r_3, 4\pi]. \tag{4.30}$$

Since $f(x)$ has five local extrema $\{s_0, s_1, s_2, s_3, s_4\}$, at least one of the intervals K_i must contain at least two consecutive extrema s_i and s_{i+1}. Let K_j be this interval. Since p is continuous and has no roots in the interior of K_j, $p(x)$ has constant sign in the interior of K_j. Without loss of generality, suppose $p(x) \geq 0$ for $x \in K_j$. However, since K_j must contain two consecutive extrema s_i and s_{i+1} of $f(x)$, we have $f(x) = 1$ for one these two and $f(x) = -1$ for the other. Without loss of generality, suppose $f(s_i) = -1$, so recalling that we assume $p(s_i) > 0$, we conclude that (4.29) does not hold for $x = s_i$ and $\varepsilon = 1/2$. This is a contradiction, so we conclude that $f(x) = \cos(x)$ cannot be approximated by a cubic polynomial on the interval K.

We also observe that fixing any degree $d \geq 1$, we can use the same argument on a sufficiently large interval K to show that $\cos(x)$ cannot be approximated in the sense of (4.29) by polynomials of degree d. Alternatively, we could fix the interval and show that $\cos(kx)$ for k large enough cannot be approximated by polynomials of degree d.

4.4 Why is non-linearity in ANNs necessary?

We now explain why in our case study of classification of handwritten digits, the exact classifier is taken non-linear. Any attempt to approximate the exact classifier by a linear function will usually lead to gross misclassifications; one cannot approximate a curve arbitrarily close by a straight line.

4.4.1 0 + 0 = 8?

This example considers the simplified setting of a DNN with no biases. That is, we use linear maps in place of affine, and we do not have any activation function, so the resulting DNN is linear. There is also no softmax layer, but it can be easily added without changing the result.

Consider the problem of classifying handwritten digits. The exact classifier $\hat{\phi}$ maps each image $\mathbf{s}$ of a handwritten digit to the integer between 0 and 9 depicted in the image. Consider the images of digits $\mathbf{s}_1$ and $\mathbf{s}_2$ in Fig. 4.5. Both images depict the digit zero. By adding these two images pixel-by-pixel, we obtain a new image $\mathbf{s}_1 + \mathbf{s}_2$, which looks very much like an eight. This leads to the provocative name of this example, "0 + 0 = 8" [49]. Observe that $\mathbf{s}_1 + \mathbf{s}_2$ is classified as 8:

$$\hat{\phi}(\mathbf{s}_1 + \mathbf{s}_2) = 8, \tag{4.31}$$

whereas

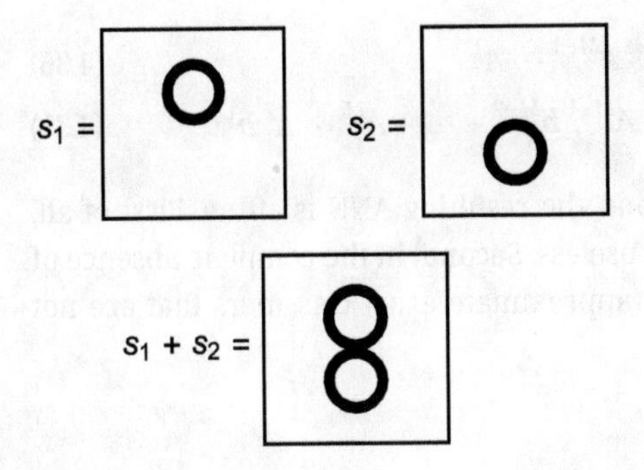

Figure 4.5: Adding together two images of a zero gives an image of an eight.

$$\hat{\phi}(\boldsymbol{s}_1) + \hat{\phi}(\boldsymbol{s}_2) = 0 + 0 = 0. \tag{4.32}$$

That is, the sum of the classes is not equal to the class of the sum. Since $\hat{\phi}(\boldsymbol{s}_1) + \hat{\phi}(\boldsymbol{s}_2) \neq \hat{\phi}(\boldsymbol{s}_1 + \boldsymbol{s}_2)$, $\hat{\phi}$ is non-linear, so it cannot be closely approximated by a linear ANN without an activation function.

When attempting tasks that are more complex than recognizing digits, affine classifiers may make even less sense. To quote directly from [49], for example, "recognizing faces instead of numbers requires a great many pixels — and the input–output rule is nowhere near linear."

4.4.2 Non-linear activation functions are necessary in ANNs

Suppose that we do not use a non-linear activation function λ in a single neuron function. That is, each neuron function (4.1) in our network has the simplified affine form

$$f(\boldsymbol{x}) = \boldsymbol{\alpha} \cdot \boldsymbol{x} + \beta \tag{4.33}$$

for a vector of parameters $\boldsymbol{\alpha}$. Then a layer function (4.3) has the simplified form

$$g(\boldsymbol{x}) = A\boldsymbol{x} + \boldsymbol{b} \tag{4.34}$$

for a matrix of parameters A and a bias vector $\boldsymbol{b}$. An ANN (4.8) composed of several linear layers has the form

$$h(\boldsymbol{x}) = (\boldsymbol{b}^M + A^M(\boldsymbol{b}^{M-1} + A^{M-1} \cdots (\boldsymbol{b}^1 + A^1\boldsymbol{x}) \cdots)), \tag{4.35}$$

where each A^i $(i = 1, \ldots, M)$ is a matrix of parameters and each $\boldsymbol{b}^i$ is a vector. However, the product of several matrices is still a matrix. Therefore, an ANN which is a composition of several affine layers is actually a single layer function:

$$h(\boldsymbol{x}) = A\boldsymbol{x} + \boldsymbol{b}, \quad \text{where } A = A^M A^{M-1} \cdots A^1 \tag{4.36}$$

$$\text{and} \quad \boldsymbol{b} = \boldsymbol{b}^M + A^M \boldsymbol{b}^{M-1} + A^M A^{M-1} \boldsymbol{b}^{M-2} + \cdots + A^M \cdots A^2 \boldsymbol{b}^1. \tag{4.37}$$

Thus, without non-linear activation functions, the resulting ANN is affine. First of all, this renders all intermediary hidden layers useless. Second, in the complete absence of non-linear activation functions, we cannot approximate exact classifiers that are not affine.

4.5 Exercises

Exercise 1 (A basic example of classifiers). Assume that we have data points in the interval $[0,1] \subset \mathbb{R}$ belonging to three possible classes. We are looking for possible soft classifiers.

a. Recall why such classifiers are functions $p : \mathbb{R} \to \mathbb{R}^3$.

b. Show that the following two functions are acceptable classifiers:

$$p(x) = \left(\frac{x}{2}, \frac{x^2}{2}, 1 - \frac{x}{2} - \frac{x^2}{2}\right),$$
$$\tilde{p}(x) = (1 - 4x(1-x) - x(1-2x)^2, 4x(1-x), x(1-2x)^2).$$

c. Is the function

$$q(x) = (x, 1-x, (x-1/2)^2)$$

an acceptable classifier?

d. Consider the following data points in $\mathbb{R}$: $x_1 = 0$ in class #1, $x_2 = 1/2$ in class #2, and $x_3 = 1$ in class #3. Calculate the results of the two classifiers p and $\tilde{p}$ on those data points. Which of the two has the best performance?

Exercise 2 (Some examples with the softmax function).

a. Apply the softmax function to the vector

$$y = (1, 0, -1) \in \mathbb{R}^3.$$

b. Can one find a vector $y \in \mathbb{R}^3$ such that

$$\sigma(y) = (1, 0, 0)?$$

c. Explain why a softmax function cannot directly yield a hard classifier.

Exercise 3 (A basic example of ANN). We consider an ANN with exactly one hidden layer. All activation functions are ReLU, and there are no biases. The neurons from the input to the hidden layer are given by the matrix

$$A = \begin{bmatrix} 1 & 2 & 3 \\ 3 & 4 & 5 \end{bmatrix},$$

while the neurons from the hidden layer to the output are given by the matrix

$$B = \begin{bmatrix} 1 & 1 \\ 2 & 2 \\ 1 & 1 \\ 2 & 2 \end{bmatrix}.$$

a. Give the dimensions of the input layer, the hidden layer, and the output layer.
b. Write the explicit formula for the full network function.
c. Give the general condition on two matrices A and B so that there exists an ANN with one hidden layer for which $A = A^1$ is the matrix for the neuron functions between the input and the hidden layer and $B = A^2$ is the matrix for the neuron functions between the hidden layer and the output.

Exercise 4 (Representing non-linear functions with ANNs). We only consider in this exercise ANNs with ReLU as activation function for all neurons.

a. Construct a neural network with a 1D input layer, a 1D output layer, and exactly one hidden layer such that the network function is exactly

$$f : x \in \mathbb{R} \to f(x) = |x| \in \mathbb{R}.$$

What is the minimal dimension of the hidden layer?

b. Extend the previously constructed network to obtain the network function

$$f : x \in \mathbb{R}^n \to f(x) = (|x_1|, |x_2|, \dots, |x_n|) \in \mathbb{R}^n.$$

c. We now want to prove that it is not possible to obtain exactly the network function, even with an arbitrary number of hidden layers

$$f : x \in \mathbb{R} \to f(x) = x^2 \in \mathbb{R}.$$

i. Assume that all biases in the network are 0. Prove that in this case, the network function f necessarily satisfies

$$f(\lambda x) = |\lambda| f(x), \quad \text{for all } \lambda \in \mathbb{R},$$

so that $f(x) \neq x^2$.

ii. In the general case with biases, prove that a network function is necessarily piecewise affine.

d. Can any piecewise linear function on $\mathbb{R}$ be obtained as a network function?

Exercise 5 (A simple classification problem). We consider a set Ω of objects, i. e., points of $\mathbb{R}^n$, that are in two different classes or subsets Ω_1 and Ω_2. We assume that Ω_1 and Ω_2 are separated by a hyperplane: there exist $\alpha \in \mathbb{R}^n$ and $\beta \in \mathbb{R}$ such that

$$x \in \Omega_1 \implies \alpha \cdot x + \beta < 0, \quad x \in \Omega_2 \implies \alpha \cdot x + \beta > 0.$$

a. Construct an ANN with input of dimension n and output of dimension 2 such that the network function $f(x) = (f_1(x), f_2(x))$ classifies Ω_1 and Ω_2 in the following sense:

$$x \in \Omega_1 \implies f_1(x) > f_2(x), \quad x \in \Omega_1 \implies f_1(x) > f_2(x).$$

b. Extend the network constructed above to the case of four classes Ω_i, $i = 1 \dots 4$, separated by two non-parallel hyperplanes: there exist $\alpha, \tilde{\alpha} \in \mathbb{R}^n$, and $\beta, \tilde{\beta} \in \mathbb{R}$, such that α and $\tilde{\alpha}$ are not collinear and

$$\begin{aligned}
x \in \Omega_1 &\implies \alpha \cdot x + \beta < 0 \text{ and } \tilde{\alpha} \cdot x + \tilde{\beta} < 0, \\
x \in \Omega_2 &\implies \alpha \cdot x + \beta > 0 \text{ and } \tilde{\alpha} \cdot x + \tilde{\beta} < 0, \\
x \in \Omega_3 &\implies \alpha \cdot x + \beta < 0 \text{ and } \tilde{\alpha} \cdot x + \tilde{\beta} > 0, \\
x \in \Omega_4 &\implies \alpha \cdot x + \beta > 0 \text{ and } \tilde{\alpha} \cdot x + \tilde{\beta} > 0.
\end{aligned}$$

5 Supervised, unsupervised, and semisupervised learning

5.1 Basic definitions

The training of machine learning algorithms is commonly subdivided into so-called supervised, unsupervised, and semisupervised learning. While those notions are often intuitively clear, there does not appear to be a commonly accepted agreement on the definitions of supervised and unsupervised learning in the literature. For example, we quote [17], where it is stated that "Unsupervised learning and supervised learning are not formally defined terms. The lines between them are often blurred."

The concept of supervised, unsupervised, or semisupervised learning was nevertheless recognized early on as useful when considering how the available datasets impact the choice of the algorithms; see [8, 18] or again the more recent [11, 17].

To compare the ideas of *supervised learning* and *unsupervised learning*, we focus on classification and clustering problems where more precise definitions can be given. In both cases, objects must be assigned to different classes. But roughly speaking, supervised learning deals with problems where the classes of the object in the training set are known a priori; we will then say that the training set is labeled. In unsupervised learning, the classes of the objects in the training set are not known a priori and the training set is said to be unlabeled.

To go back to our example of handwriting recognition, images of handwritten digits need to be placed into one of ten classes, or images of handwritten English letters have to be placed into one of 26 classes. When applying supervised learning, we require a labeled training set. The labels (that is, classes) are for example "0," "1," . . . , "9" for digits. Our training set is a collection T of images $\boldsymbol{s}$ together with the label corresponding to each image.

When training a machine learning algorithm in this supervised setting, one can compare the outputs from the algorithm to the labels that are known to be correct. When the algorithm's output does not match the correct label, it improves itself ("learns") and will hopefully produce more accurate outputs in the future. Once training is over, the parameters of the algorithm are fixed and this algorithm can then be used on a wider range of images outside of T.

In contrast, *unsupervised learning* algorithms learn from unlabeled training sets. In the previous example, that would mean that our training set contains only a collection of images without knowing to which digit each image corresponds. Instead of classification, one sometimes refers to this as a *clustering problem*. Like for classification problems, in a clustering problem, one has a set S of objects in $\mathbb{R}^n$ and seeks to group together objects in S which are similar in some way. In some cases, we may still know how many groups to form. For example, if we know that we have images of handwritten digits, we also know that we should obtain ten groups; but in other situations, even

https://doi.org/10.1515/9783111025551-005

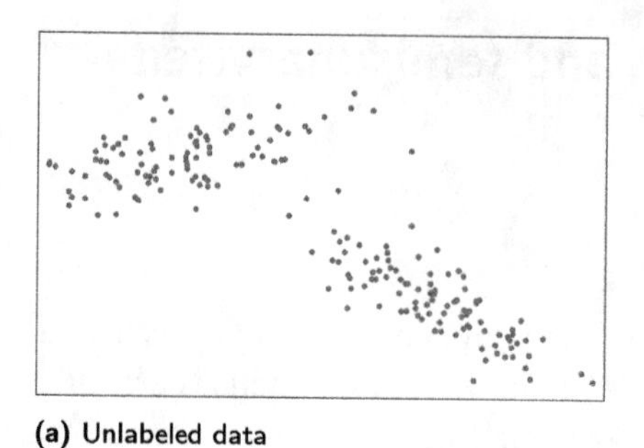

(a) Unlabeled data

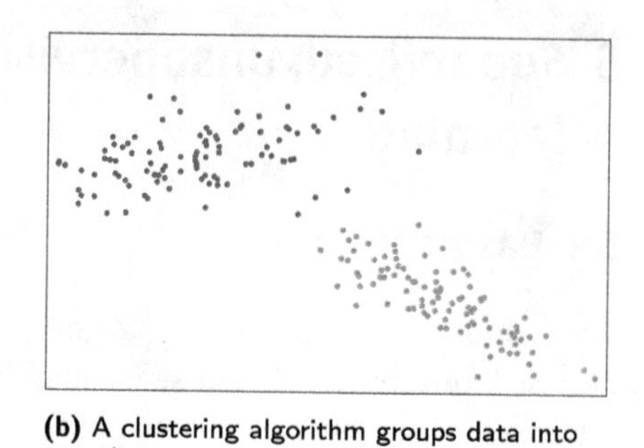

(b) A clustering algorithm groups data into two clusters

Figure 5.1: Result of a clustering algorithm in unsupervised learning.

the number of classes is unknown. Figure 5.1 depicts an example of a simple clustering problem where we start from unlabeled data (left) and end up with two classes (right). We next present two simple examples of unsupervised clustering where the classes are not known a priori.

Example 5.1.1. Suppose we are trying to translate some writing from an unknown ancient language. Since there is no dictionary, the words are unknown. We can only group pieces of writing which may be words by similarity. That is, we can identify clusters of similar objects and assume that they are words in this ancient language.

The following example presents a realistic application of unsupervised learning.

Example 5.1.2. Suppose we visit a tree farm and want to classify the different species of tree that grows there. Before our trip, we were told how many species of trees grow at the farm but nothing more. To classify the trees, we collect a sample of leaves from every tree from the farm for measurement. For each leaf, we can determine the color, size, and shape. With these data, we can group similar leaves together since leaves from one species of tree should be similar regardless of which tree the leaf was taken from (i. e., leaves from an oak tree never look like leaves from a pine tree). In this way, we can cluster similar leaves together and assume that each cluster represents one species of tree.

The architecture of ANNs in unsupervised learning is necessarily somewhat different from that of ANNs in supervised learning. Notably, one cannot simply compare the output of the algorithm with the known label anymore. This observation implies that unsupervised learning requires an ANN structure that differs from supervised learning. For example, simply adding more layers often widens the class of classifier functions that an ANN can approximate, but such a simple strategy does not work for unsupervised learning. As this is an introductory book, we will not explain how ANNs may be used for unsupervised learning and simply mention the use of so-called auto-encoders or the popular self-organizing maps; see, e. g., [24].

Recall that our set of objects S is divided into mutually disjoint classes, $S = \bigcup_{j=1}^{m} S_j$, and the exact classifier $\hat{\phi}$ maps objects to the index of their class: $s \mapsto j$. The training set is a subset, $T \subset S$, on which the machine learning algorithm is trained to fit its parameters, such as the weights and biases of an ANN in Definition 4.1.3.

This intuitive discussion leads to the following rigorous definitions adopted in our lectures.

Definition 5.1.1. A machine learning algorithm is called:
- *supervised* if the exact classifier $\hat{\phi}$ is known on the whole training set T (all data in T are labeled);
- *unsupervised* if the exact classifier is completely unknown on any object $s \in T$ (no data in T are labeled);
- *semisupervised* if the exact classifier is known on some strict subset $T_0 \subset T$ (some data in T are labeled).

In general the decision to use supervised, unsupervised, or semisupervised algorithms depends on, for example, the availability of appropriate datasets. For instance, if one wishes to use supervised learning on some datasets, it may be necessary to manually classify every object in the training set. This may be very time consuming or simply impossible to achieve for large and "messy" training sets. In such circumstances, the use of unsupervised or semisupervised algorithms may be required.

We also emphasize that unsupervised learning is sometimes useful for making inferences about and identifying patterns from only the training set T, while a supervised algorithm is always expected to be used on a larger set of objects.

5.2 Example of unsupervised learning: detecting bank fraud

Consider the following example of a problem which employs unsupervised learning. Banks have an interest in determining whether or not transactions made via their accounts are fraudulent. This is an important practical problem, and the solutions banks implement are quite sophisticated. Moreover, banks keep the details of their procedures proprietary to make their security measures harder to circumvent. Therefore, we present here a simple example which illustrates how the problem of bank fraud could hypothetically be addressed by unsupervised learning.

The average account holder makes several transactions every day, and each transaction has several quantifiable properties which can be used to represent each transaction as a point in $\mathbb{R}^n$. For example, a transaction could be represented in $\mathbb{R}^4$ as

$$s = \begin{pmatrix} \text{transaction amount} \\ \text{transaction time} \\ \text{day of the week} \\ \text{method, e. g., 0 for in-person, 1 for online} \end{pmatrix}. \tag{5.1}$$

An unsupervised clustering algorithm trained on the set of a bank account holder's transactions would find that most account holders have many similar transactions – they tend to spend similar amounts at similar stores and websites at similar times of the day and week, leading to clusters of transactions that are all comparable. Therefore, a transaction that does not lie in a cluster with many of the account holder's other transactions is more likely to be fraudulent and warrants closer attention.

5.3 Exercises

Exercise 1 (Basic comparison of supervised vs. unsupervised learning). Consider the following points in $\mathbb{R}$: $x_1 = -1$, $x_2 = 1$, and $x_3 = a$ for some $-1 < a < 1$.

a. Assume we know that x_2 and x_3 belong to the same class, while x_1 belongs to a different class. Give an example of a classifier that fits the data.

b. Assume instead that we do not know the classes; we only know that there are exactly two classes so that we instead try to cluster the points x_1, x_2, x_3 into two groups. Explain which is the most logical clustering depending on the value of a.

Exercise 2 (Basic example of k-means algorithm in dimension 1). The k-means algorithm is a very well-known algorithm to perform clustering tasks in the context of unsupervised learning.

Consider the simplest case where we are given n data points in $\mathbb{R}$: $x_1, x_2, \ldots, x_n$. We wish to cluster those points into two separate classes C_1 and C_2.

a. Recall the definition of the variance of the points $x_1, \ldots, x_n$.

b. Show that this variance can also be expressed as

$$\operatorname{Var}(x_i)_{i\in\{1,\ldots,n\}} = \operatorname{Var}(x_1,\ldots,x_n) = \frac{1}{2N}\sum_{i,j=1}^{n} |x_i - x_j|^2 = \sum_{i=1}^{N} x_i^2 - \frac{1}{N}\left(\sum_{i=1}^{N} x_i\right)^2.$$

c. We now divide the indices $i = 1, \ldots, N$ into two subsets C_1 and C_2 with $C_1 \cap C_2 = \emptyset$ and $C_1 \cup C_2 = \{1, \ldots, N\}$. Give one explicit expression for the following function which is the sum of the variances over each subset:

$$L(C_1, C_2) = \operatorname{Var}(x_i)_{i\in C_1} + \operatorname{Var}(x_i)_{i\in C_2}.$$

d. Show that we can write

$$L(C_1, C_2) = \sum_{i=1}^{n} |x_i|^2 - \frac{1}{N}\left(\sum_{i\in C_1} x_i\right)^2 - \frac{1}{N}\left(\sum_{i\in C_2} x_i\right)^2.$$

e. We denote

$$x = \sum_{i \in C_1} x_i - \frac{1}{2} \sum_{i=1}^{n} x_i = \frac{1}{2} \sum_{i \in C_1} x_i - \frac{1}{2} \sum_{i \in C_2} x_i.$$

Show that there exists a constant $\bar{L}$ such that

$$L(C_1, C_2) = \bar{L} - \frac{2}{N} x^2.$$

f. The goal of the k-means clustering algorithm is to choose C_1 and C_2 so as to minimize $L(C_1, C_2)$. Explain what is the best choice in terms of x.

g. In general, finding the optimal C_1, C_2 corresponding to the optimal x is a complex algorithmic problem. But in dimension 1, it is possible to considerably simplify this, even if k-means clustering is not especially useful in dimension 1. First we order x_i as $x_1 < x_2 < \cdots < x_n$. Show that the optimal C_1 and C_2 are necessarily of the type

$$C_1 = \{i \leq i_0\},\ C_2 = \{i > i_0\} \quad \text{for some } i_0, \quad \text{or}$$
$$C_1 = \{i > i_0\},\ C_2 = \{i \leq i_0\} \quad \text{for some } i_0.$$

6 The regression problem

6.1 What is regression? How does it relate to ANNs?

Regression analysis is a well-known and widely used tool in data science and statistics dating back to the nineteenth century. The term *regression* was coined in [14]; see also [54]. Roughly speaking, the goal of regression analysis is to find a function of some independent variables (called predictors) that best fits the given data; see, e. g., [37]. For example, with one predictor, the aim of regression analysis would be to find a curve (called a *regression curve*) which provides the best fit to some discrete empirical data, such as a finite set of points in a 2D plane; see Figs. 6.1 and 6.2. In contrast to the well-known problem of interpolation of discrete values of a function, in regression, the desired function does not have to pass through each point, but the goal is to minimize some error, for example with the method of least squares; see, e. g., [48] for a description of interpolation.

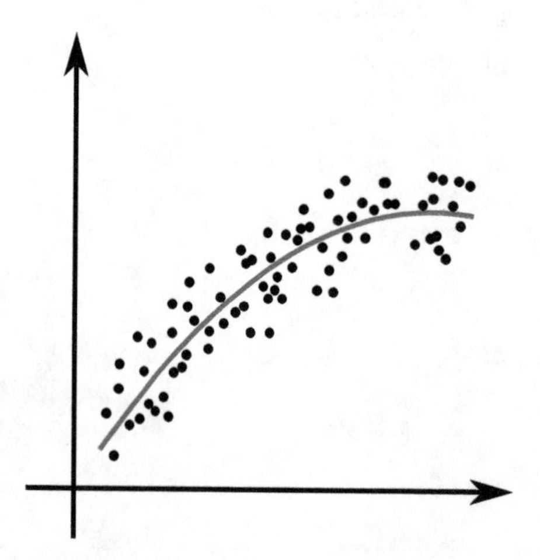

Figure 6.1: An simple example of regression. Black dots represent data points from measurements, and a regression curve is shown in red.

A well-known case concerns linear regression where one is trying to find the "best" linear combination of predictors. Geometrically, this linear combination $c_1x_1 + \cdots + c_{n-1}x_{n-1} + b = y$ is a hyperplane where x_i are the predictors and c_i are the coefficients of the linear combination. Again in the case of one predictor, the hyperplane (or linear space) is simply a line, $cx + b = y$.

A typical regression problem can be rigorously formulated as follows. One seeks to approximate an unknown function $f : \mathbb{R}^n \to \mathbb{R}^m$. In the simplest setting, the value of $f(x)$ is assumed to be known at the sample points $\mathbf{x}_1, \ldots, \mathbf{x}_N$. That is, one has a finite set of pairs $(\mathbf{x}_i, \mathbf{y}_i)$ which are data points in the set

https://doi.org/10.1515/9783111025551-006

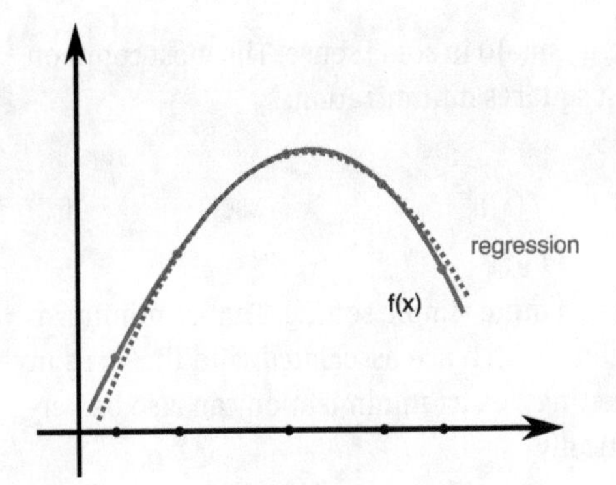

Figure 6.2: The red points are data sampled from the gray curve $f(x)$. The blue dotted line is the regression curve.

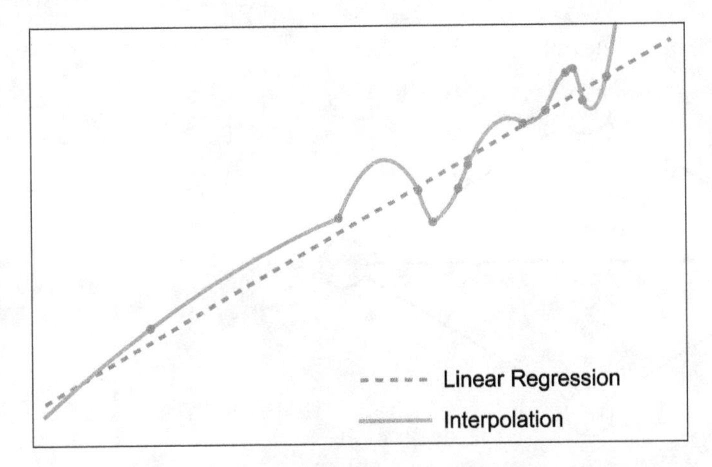

Figure 6.3: Linear regression vs. interpolation. The interpolated curve passes through every data point while the linear regression function does not pass through most of the points. Linear regression captures the trend of the data while interpolation is more sensitive to noise, for example due to measurement errors.

$$T = \{(\boldsymbol{x}_1, \boldsymbol{y}_1), \ldots, (\boldsymbol{x}_N, \boldsymbol{y}_N)\} \subset \mathbb{R}^n \times \mathbb{R}^m \tag{6.1}$$

such that $f(\boldsymbol{x}_i) = \boldsymbol{y}_i$ for each $i = 1, \ldots, N$.

However, in many circumstances, we may expect some kind of error or "noise" in the dataset T, for example, measurement errors in y_i or x_i. In those settings, we are not expecting to have $f(x_i) = y_i$ for the "actual" function f. This offers another example where pure interpolation may not be the optimal strategy and a best fit from regression may actually be closer to the objective function f, see Fig. 6.3.

One seeks an approximation of the function f in the form of $\tilde{f}(x, \gamma)$, where γ is a vector of tunable parameters. The regression problem consists of optimizing the parameters γ so that $f(x) - \tilde{f}(x,\gamma)$ is optimal (e. g., small) in some sense. The most common example of such optimization is simply least squares minimization:

$$\min_{\gamma \in \mathbb{R}^{\mu}} \sum_{i=1}^{N} \|\tilde{f}(x_i,\gamma) - f(x_i)\|^2. \tag{6.2}$$

This optimization is an example of training in a quite simple setting. That is, minimization in (6.2) can be performed in iterative steps which are associated with the steps in the learning process. In some very simple settings, exact minimization can also be performed and results can be calculated analytically.

In general, regression problems are also a natural framework for deep neural networks (DNNs). Simply put, we consider a function $\tilde{f}(x,\gamma)$ that is an ANN function $h(x,\boldsymbol{\gamma})$ as in Definition 4.1.3; $\boldsymbol{\gamma}$ is the collection of parameters, e. g., weight matrices, see Fig. 6.4.

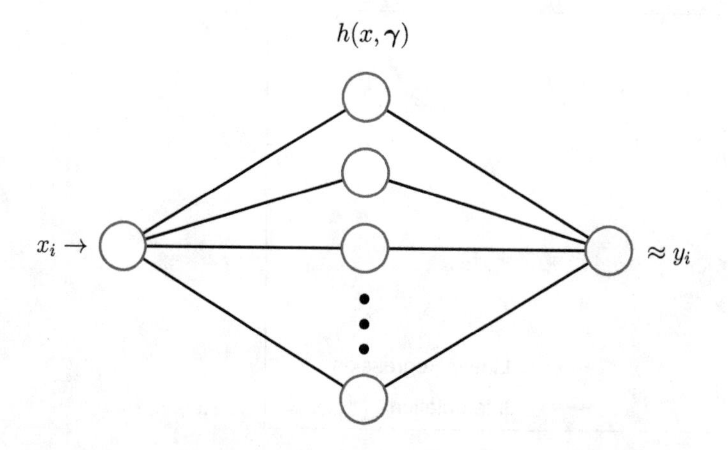

Figure 6.4: Regression ANN with input, one output, and one hidden layer.

6.2 Example: linear regression in dimension 1

The simplest case of a regression problem is *linear regression* to approximate a function $f : \mathbb{R} \to \mathbb{R}$ by a linear function $\tilde{f}(x,\gamma)$, i. e., $\tilde{f}(x,\gamma) = ax + b$ with unknown parameters $\gamma = (a,b) \in \mathbb{R}^2$.

In many settings, the optimization of parameters through training is typically a complicated, iterative process. In contrast, here we are lucky because the optimization problem (6.2) to find a and b can be solved analytically using calculus, and the resulting formulas are given in terms of data points (x_i, y_i) from (6.1):

$$a = \frac{N\sum_{i=1}^{N} x_i y_i - (\sum_{i=1}^{N} x_i)(\sum_{i=1}^{N} y_i)}{N\sum_{i=1}^{N} x_i^2 - (\sum_{i=1}^{N} x_i)^2}, \quad b = \frac{\sum_{i=1}^{N} y_i - a\sum_{i=1}^{N} x_i}{N}. \tag{6.3}$$

Thus, the optimal parameters for linear regression can be found in a single simple step in this elementary case.

6.3 Logistic regression as a single neuron ANN

As explained above, the goal of regression analysis is to find a curve that fits the given data. The family of such curves with unknown parameters is called a *regression model*. In particular, in the simple case of linear regression in 2D the curve is a straight line which provides the best fit for pairs of data points on the plane.

In contrast, a *logistic model* [9], also known as a logit model, can describe the probability of events with binary outcomes such as pass/fail in exams or positive/negative in COVID-19 tests. It is possible to generalize this model for more than two outcomes such as the classification of ten handwritten digits, but we will focus here on binary outcomes. Given data and a logistic model that is a parameterized family of probability distributions, logistic regression aims to find the parameters that provide some kind of best fit (e. g., minimizing cross-entropy loss, described below) between the data and the probability distribution. For instance, in the example in Fig. 6.6 the probability distribution provides a higher probability of passing for the people who actually passed and a low probability of passing for the people who actually failed.

One can compare regression with logistic regression in the following way: regression aims to minimize the distance between the data and the approximating curve, while logistic regression aims to obtain a probability distribution which maximizes the probability of observing the given data.

The notion of *logistic regression* [10] is commonly used to analyze binary logistic models. Typically, one approximates a function $p(\mathbf{x})$ which gives the probability that an event occurs based on the value of some independent variable $\mathbf{x}$.

We specifically discuss the relations between logistic regression and two areas of machine learning (support vector machines (SVMs) and DNNs) through examples.

6.3.1 1D example: studying for an exam

Consider a group of students preparing for an exam. Each student studies a certain number of hours. Students who study more are clearly more likely to pass the exam than students who study less, though studying for a long time is not a guarantee of passing, and neither is not studying a guarantee of failure. We assume that this situation can be accurately described by a random process, namely, there exists a function $p(x)$ that corresponds to the probability of passing the exam after x hours of study. We can expect $p(x)$ to increase from near 0 when $x = 0$ (students who do not study are unlikely to pass) to near 1 for large x (students who study a lot are very likely to pass). See Fig. 6.6 for an example of the curve $p(x)$. A logistic regression algorithm seeks to approximate this function.

The name "logistic" refers to the form of the approximation $\tilde{p}(x, \gamma)$ of $p(x)$ that we are trying to obtain:

$$\tilde{p}(x, \gamma) = \frac{1}{1 + e^{-(ax+b)}}, \quad x \in \mathbb{R}, \tag{6.4}$$

where $\gamma = (a, b) \in \mathbb{R}^2$. The function

$$\sigma(x) = 1/(1 + e^{-x}) \tag{6.5}$$

is called the *logistic function* (which gives the name to logistic regression) or *sigmoid function*. It is often used as an activation function in ANNs, so $\tilde{p}(x, \gamma) = \sigma(ax + b)$ is a composition of a non-linear (activation) function with an affine function, just as in a single neuron function in Definition 4.1.1.

In Fig. 6.5 we see that $\sigma(x)$ is monotone increasing from 0 as $x \to -\infty$ to 1 as $x \to \infty$. Therefore, it is a good simple approximation of probabilities evolving from 0 at $-\infty$ (probability of an event that never occurs) to 1 at ∞ (probability of an event that almost surely occurs). The two parameters a and b of the affine transform in (6.4) control the slope of $\tilde{p}(x, \gamma)$ and the location of the point where $\tilde{p}(x, \gamma) = 1/2$. But of course, we cannot expect to fit any such probability function $p(x)$ with only the function $\sigma(x)$ of the form (6.5).

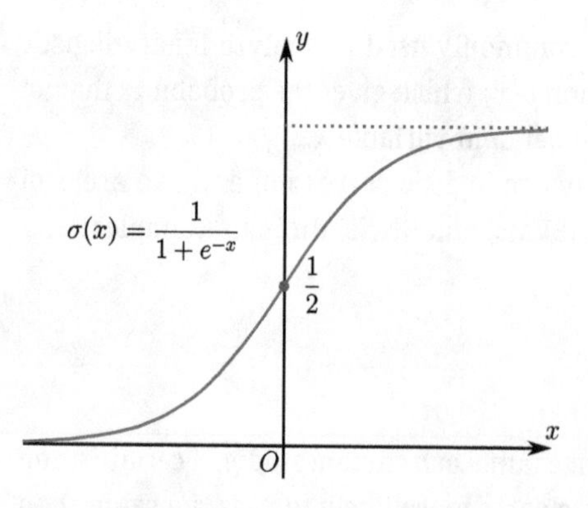

Figure 6.5: Sigmoid red curve $\sigma(x)$ – note that it asymptotically approaches the dotted line $y = 1$ as $x \to \infty$ and approaches zero as $x \to -\infty$.

Recall that in the basic regression problem described in Section 6.1 (e. g., (6.1)–(6.2)), one knows several exact values taken by the unknown function to be approximated (see (6.1)). In contrast, in the above example of logistic regression, one does not know the exact value of $p(x)$ at any point since we cannot measure the probability of any student passing an exam.

Instead, we have the following data from N students: pairs (x_i, r_i), $i = 1, \ldots, N$, where $x_i \geq 0$ is the number of hours the student spent studying and r_i is 0 if the student failed and 1 if they passed. We do not expect to be able to predict with perfect accuracy whether any given student passes or fails based only on the number of hours studied because there are many other factors at play, e. g., attendance. Instead we will use these data to find the optimal parameters $\gamma = (a, b)$ in the logistic regression function $\tilde{p}(x, \gamma)$ defined in (6.4) to approximate the probability that a student passes the exam.

First, we can arrange the known data pairs (x_i, r_i) so that $r_i = 1$ for $i = 1, \ldots, k < N$ and $r_i = 0$ for $i = k+1, \ldots, N$, i. e., the first k students passed the exam and the last $N - k$ students failed. Next, we can for example minimize the following function:

$$L(\gamma) := -\sum_{i=1}^{k} \log(\tilde{p}(x_i, \gamma)) - \sum_{i=k+1}^{N} \log(1 - \tilde{p}(x_i, \gamma)). \tag{6.6}$$

The function $L(\gamma)$ in (6.6) is called the *cross-entropy loss function*. The minimization of this function over all parameters γ is a typical example of a training process which is also used in DNN classifiers.

Since $\tilde{p}(x, \gamma)$ has the logistic form (6.4), we arrive at minimizing the following function over all $(a, b) \in \mathbb{R}^2$:

$$L(a, b) := -\sum_{i=1}^{k} \log\left(\frac{1}{1 + e^{-(ax_i+b)}}\right) - \sum_{i=k+1}^{N} \log\left(\frac{e^{-(ax_i+b)}}{1 + e^{-(ax_i+b)}}\right). \tag{6.7}$$

An example with specified numbers (x_i, r_i) of the optimal logistic regression $\tilde{p}(x, \gamma)$ for the exam studying example is shown in Fig. 6.6.

In general, of course, the solution to the optimization problem can be intricate. However, there are cases where the solution is simple and understandable, e. g., when we have a steep transition between a probability of succeeding from almost 0 to almost 1. Hence assume that $p(x) \approx 0$ for $x < \bar{x}$ and $p(x) \approx 1$ for $x > \bar{x}$ (here "$\approx$" stands for "is close to"). In that case, in the first sum in (6.7) over students who passed, we expect most of them to have had a high probability of passing, i. e., $p(x_i) \approx 1$. Therefore, we need to find γ so that the logistic regression $\tilde{p}(x_i, \gamma)$ is also close to 1. That is, we want $\log(\tilde{p}(x_i, \gamma)) \approx 0$ or $a\, x_i + b \gg 1$ for $x_i > \bar{x}$ (here "$\gg$" stands for "is much greater than"). Similarly, in the second sum for failing students, we would need $1 - \tilde{p}(x_i, \gamma) \approx 1$ or equivalently $a\, x_i + b \ll -1$ for $x_i < \bar{x}$. This naturally leads to $a \gg 1$ and $b = -a\,\bar{x}$ so that $a\, x_i + b = a\,(x_i - \bar{x})$.

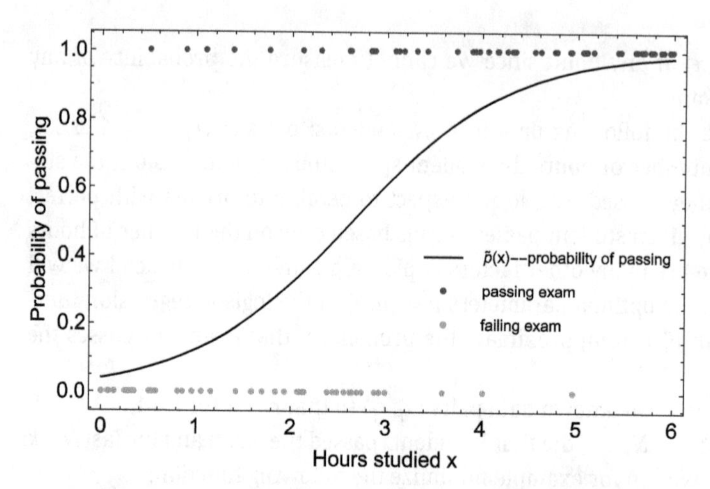

Figure 6.6: In this example, 100 students study for an exam for varying amounts of time (between 0 and 6 hours). Some students pass (blue dots) and others fail (orange dots). The black curve is the composition of the sigmoid with an affine $ax + b$ function (see the definition of a single neuron in Definition 4.1.1). It shows the function $\tilde{p}(x)$ of the form (6.4) which minimizes (6.7). We interpret $\tilde{p}(x)$ as the probability that a student passes the exam after studying for x hours.

6.3.2 2D example of admittance to graduate school: separation of sets and decision boundary

Suppose we are provided with the GPAs and GRE scores of 600 applicants to a graduate program and data on whether or not each applicant was accepted. The GPAs and GRE scores are recorded as a vector (x_1, x_2) in $\mathbb{R}^2$. Each vector (x_1, x_2) is assigned the label (class) "1" if the student was admitted and "0" otherwise. See Fig. 6.7, where sets of blue and orange dots represent these two classes.

In our previous notation, the set of all possible pairs of GPA and GRE score is the set of objects $S \subset \mathbb{R}^2$. The pool of 600 applicants whose scores we know is the training set T, and it is labeled because we know whether each applicant was accepted or not. The goal is to use logistic regression to predict the probability of acceptance of future applicants, e. g., for $s \in S \setminus T$. Note that in this problem, it is possible that a future applicant has the exact same GPA and GRE score as one of the 600 applicants in T. However, that does not necessarily mean that the outcome will be the same for these two; two applicants can have exactly the same scores with one accepted and one not due to other factors such as previous research experience, strength of letters of recommendation, etc. The whole point of performing logistic regression lies in estimating the probabilities of the outcome (rather than the exact outcome) based only on a limited number of factors.

In this example, we use the following logistic regression model:

$$\tilde{p}(\boldsymbol{x}, y) = \sigma(\boldsymbol{a} \cdot \boldsymbol{x} + b), \tag{6.8}$$

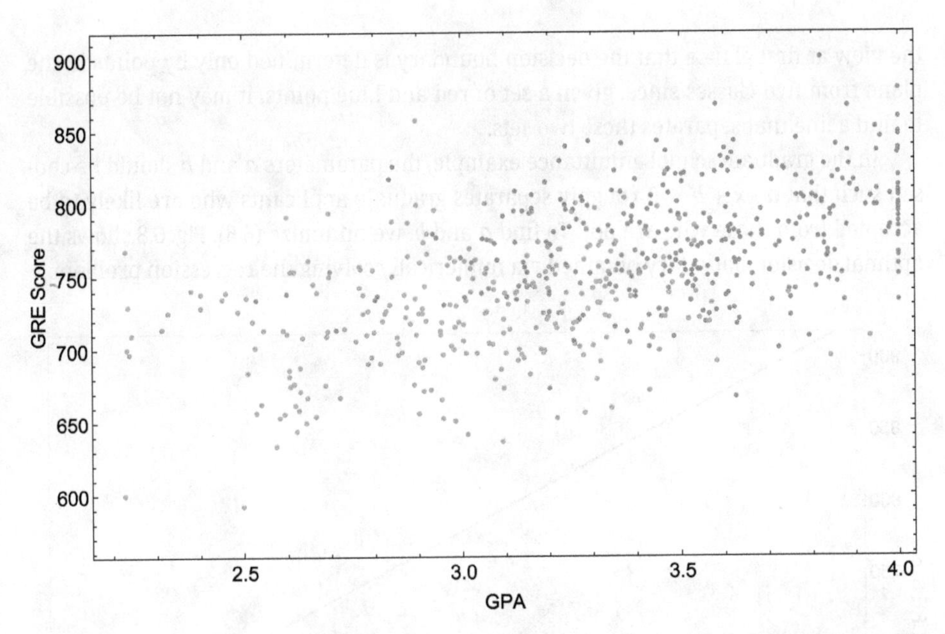

Figure 6.7: Scatter plot of GPAs and GRE scores of graduate school applicants. Blue points represent applicants who were ultimately accepted, while orange points represent applicants who were rejected.

where $\gamma = (\boldsymbol{a}, b)$, $\boldsymbol{a} \in \mathbb{R}^2$, and $b \in \mathbb{R}$. The graph of the logistic regression function (6.8) is a surface in 3D controlled by three parameters: $\mathbf{a} = (a_1, a_2)$ and b. The line $\boldsymbol{a} \cdot \boldsymbol{x} + b = 0$ is the *decision boundary* (see Fig. 6.8) for $\tilde{p}(\boldsymbol{x}, \gamma)$, with $\tilde{p}(\boldsymbol{x}, \gamma) > 1/2$ on one side of the line and $\tilde{p}(\boldsymbol{x}, \gamma) < 1/2$ on the other side.

The decision boundary does not entirely determine the logistic regression model of course. One can easily find other parameters $\tilde{\boldsymbol{a}}$ and $\tilde{b}$ yielding the same decision boundary, i. e., such that $\{\boldsymbol{a} \cdot \boldsymbol{x} + b = 0\} = \{\tilde{\boldsymbol{a}} \cdot \boldsymbol{x} + \tilde{b} = 0\}$. We can for example just take $\tilde{\boldsymbol{a}} = \lambda\, \boldsymbol{a}$ and $\tilde{b} = \lambda\, b$ for the same $\lambda \in \mathbb{R}$. In particular, that means that the decision boundary does not give any information on how steep $\tilde{p}(x, \gamma)$ is around the decision boundary.

In logistic regression as well as in SVMs described below, the decision boundary is a straight line (linear decision boundary), whereas for DNN classifiers, it could be a more complicated curve in 2D (or a surface in higher dimensions).

Ideally, all points from one class would be on one side of the decision boundary and all points from the other class would be on the other. This would lead to an exact decision boundary. However, in practical problems, the optimal decision boundary is usually a line which separates classes approximately, i. e., most points from the first class are on one side of the curve and most points from the second class are on the other side.

We emphasize that as a consequence *the decision boundary is defined by the classifier* (e. g., $\tilde{p}(\boldsymbol{x}, \gamma)$ for logistic regression) rather than the purely geometric structure of the classes. In other words, the decision boundary is determined by the classifier's decision about in which class to place each object (be it correct or not). This stands in contrast to the view at first glance that the decision boundary is determined only by points in the plane from two classes since, given a set of red and blue points, it may not be possible to find a line that separates these two sets.

In the graduate school admittance example, the parameters $\boldsymbol{a}$ and b should be chosen such that $\boldsymbol{a} \cdot \boldsymbol{x} + b = 0$ roughly separates graduate applicants who are likely to be accepted from those who are not. To find $\boldsymbol{a}$ and b, we minimize (6.6). Fig. 6.8 shows the optimal decision boundary obtained via numerically solving the regression problem.

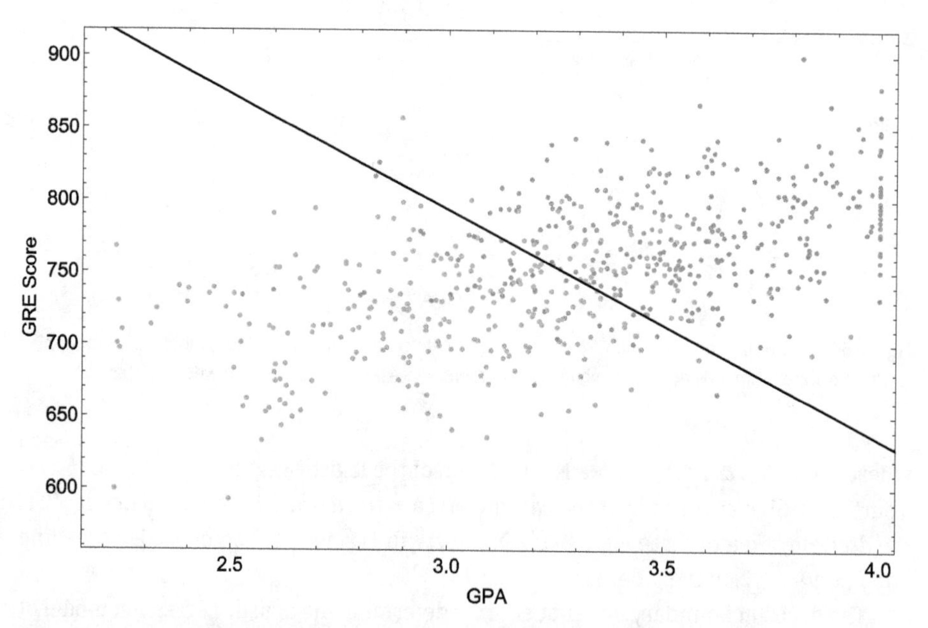

Figure 6.8: Scatter plot of GPAs and GRE scores of graduate school applicants. Blue points represent applicants who were ultimately accepted, while orange points represent applicants who were rejected. The black line is the decision boundary of the optimal logistic regression, separating applicants likely to be accepted from those likely to be rejected.

6.3.3 Relation between ANNs and regression

As mentioned above, the logistic model (6.8) is a composition of σ with an affine function $\boldsymbol{a} \cdot \boldsymbol{x} + b$. One may hence think of the iterative minimization of (6.6) in logistic regression as training an ANN with only one layer function (in fact, only one neuron function, $k = 1$ in Definition 4.1.2 of a layer function) whose activation function is the sigmoid $\sigma(x) = 1/(1 + e^{-x})$; see Fig. 6.9.

The sigmoid function is a well-known activation function, and it was even more commonly used early in the development of ANNs. As an activation function, $\sigma(x)$ models a smooth transition between an inactive neuron which is not firing ($\sigma(x) \approx 0$) and an active (firing) neuron ($\sigma(x) \approx 1$).

Exam-studying example:

(x) ——— $(\tilde{p})$ $\tilde{p}(x) = \sigma(ax + b)$

x - hour studied

Graduate admission example:

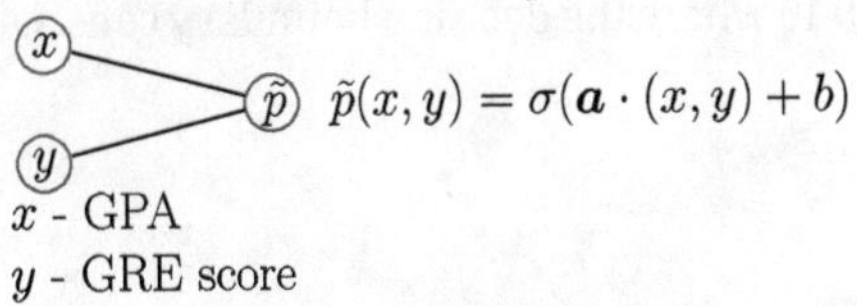

Figure 6.9: Logistic regression examples represented as ANNs with a single neuron function as in Definition 4.1.1.

In recent years, other activation functions such as ReLU have gained popularity relative to the sigmoid function; see, e. g., [16]. This is because the process of training an ANN involves iteratively taking various partial derivatives of the ANN function with respect to its parameters and using these derivatives to improve the parameter values to make better classifications (such as the gradient descent method described later). The derivative $\sigma'(x)$ becomes small for large $x > 0$, while $\mathrm{ReLU}'(x) = 1$ for large $x > 0$. If one uses derivative-based optimization methods, then the magnitude of improvement is determined by the magnitude of the derivatives, which is in turn determined by the derivative of the activation function. This means that the improvements made to ANN parameters through the training process are smaller when the activation function is σ than when it is ReLU, leading to longer training when σ is used. This behavior of sigmoid is known as the "vanishing gradient."

This leads us to summarizing the relations between regression and DNNs in the following manner.

- Linear regression introduces the concept of affine dependence on parameters, which is a key building block of ANNs.
- Logistic regression introduces a concept of activation function (i. e., sigmoid) and its composition with affine transform, which corresponds to the definition of a single layer function in DNNs.
- The optimization of parameters in logistic regression leads to the cross-entropy loss function which is often used in the training of DNNs.

6.3.4 Logistic regression vs. networks with many layers

To keep the discussion as simple as possible, let us return to our 1D example of students working on their final exam. In that setting, we see that a logistic regression function $\tilde{p}(x, y)$ is monotone increasing from 0 to 1 (or decreasing from 1 to 0 if $a < 0$). Therefore, we can only expect $\tilde{p}(x, \gamma)$ to be a good approximation of the actual (unknown) probability $p(x)$ if $p(x)$ is also monotone (or close enough to being monotone, as in Fig. 6.10).

The notion of decision boundary is useful again: Logistic regression models can be reasonably accurate only when the decision boundary where $p(x) = 1/2$ is *one point*, or "almost one point" as in the blue curve in Fig. 6.10 where the decision boundary consists of several closely spaced points.

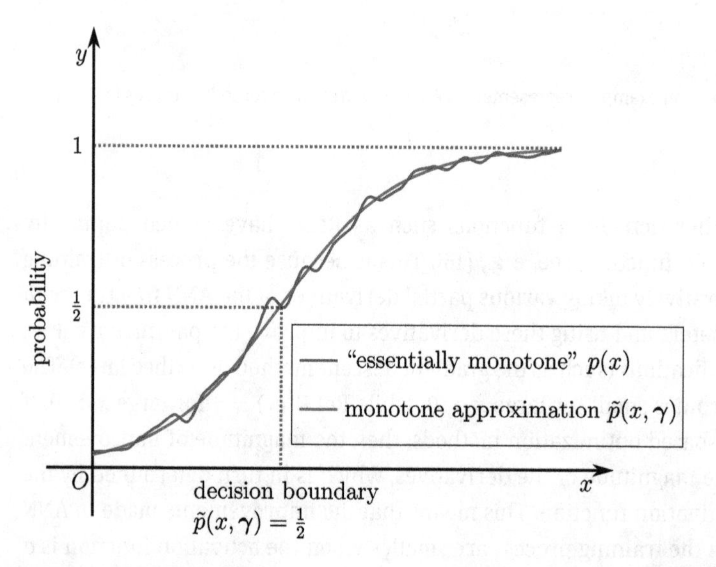

Figure 6.10: The exact probability curve $p(x)$ is not monotone, but it is essentially monotone in that $p(x) > p(y)$ whenever $x - y$ is sufficiently large. In this case, it is still possible to approximate $p(x)$ by a monotone function.

A simple example where a logistic model is likely not appropriate is the following model of growing plants. Every plant requires light of a certain intensity. If the intensity is too low, the plant is not likely to thrive. Likewise, if the light is too intense, the plant may dry out and wither as well. Thus, the probability $p(x)$ that a plant survives after a certain period of time (one week for example) under light of intensity x is not monotone; see Fig. 6.11.

If $p(x)$ represents the probability that a plant survives for a week under light of intensity x, the curve $p(x)$ initially increases from near 0 to near 1 as the light intensity changes from low to moderate and then decreases from near 1 to near 0 as the light intensity increases from moderate to high. This curve is not monotone, so it cannot be well approximated by the monotone logistic curve. Note that this curve has not one but two well-separated points where $p(x) = 1/2$.

Recall that the universal approximation theorem states that any continuous function regardless of its monotonicity can be approximated by a DNN with two or more

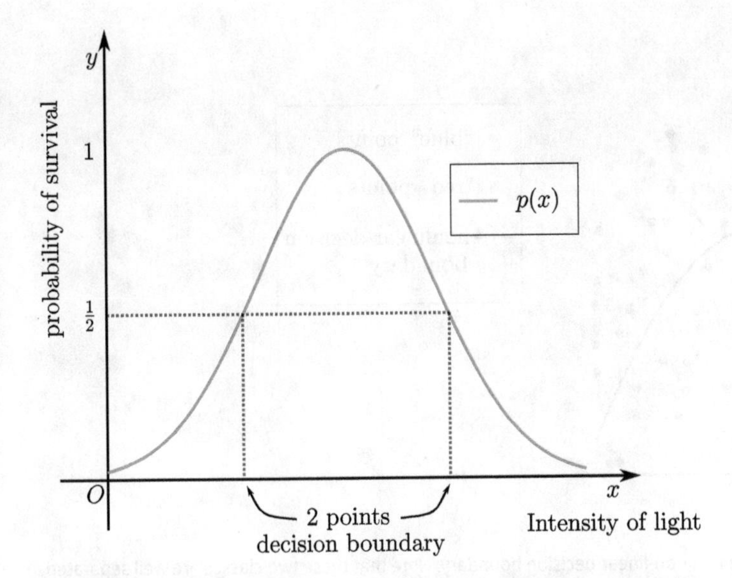

Figure 6.11: Probability distribution of a plant surviving. The points on the x-axis where $p(x) = \frac{1}{2}$ are decision boundaries. Note that this probability distribution cannot be well approximated by a sigmoid function because there is more than one decision boundary.

function layers when using the ReLU activation function. This does not apply to the logistic regression model $\tilde{p}(x, \mathbf{y})$ since it corresponds to a DNN with only one function layer and the sigmoid activation function.

In higher dimensions $n > 1$, such as in the 2D graduate school admission example, the use of logistic models similarly requires that the decision boundary where $p(\mathbf{x}) = 1/2$ is a *single* $(n-1)$-dimensional hyperplane, which is a straight line in 2D (or almost a straight line); see Fig. 6.8.

Hence in general, when the decision boundary is not almost a straight line (see Fig. 6.12) or expected to consist of several lines, the use of DNNs with at least two function layers is likely to yield a better approximation of $p(\mathbf{x})$.

6.4 Exercises

Exercise 1 (Basic example of linear regression).

a. Consider the following data points: $(x_1 = -2, y_1 = 1)$, $(x_2 = 0, y_2 = 2)$, $(x_3 = 1, y_3 = -1)$. Solve the corresponding linear regression problem, that is, find the best coefficients $a, b \in \mathbb{R}$ minimizing

$$\sum_{i=1}^{3} |a\,x_i + b - y_i|^2.$$

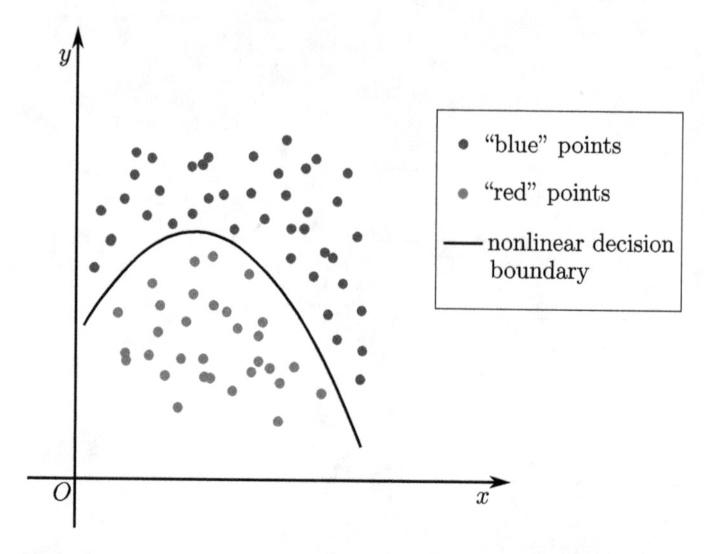

Figure 6.12: An example of a non-linear decision boundary. Note that these two classes are well separated by the curve in black (i. e., each class is on exactly one side of the curve), while any straight line decision boundary would be unable to perfectly separate the classes.

b. We now take data points x_i in $\mathbb{R}^2$ and y_i still in $\mathbb{R}$: $x_1 = (-2, 1)$, $y_1 = 1$, $x_2 = (0, -1)$, $y_2 = 2$, $x_3 = (1, 2)$, $y_3 = -1$, $x_4 = (3, 0)$, $y_4 = 0$. Solve the linear regression problem by finding the coefficients $a \in \mathbb{R}^2$ and $b \in \mathbb{R}$ minimizing

$$\sum_{i=1}^{4} |a \cdot x_i + b - y_i|^2,$$

where $a \cdot x_i$ is the usual inner product on $\mathbb{R}^2$.

Exercise 2 (Logistic regression in higher dimensions). In this chapter, we showed formulas for logistic regression functions $\tilde{p}(x, y)$, where x is 1D and 2D.

a. Write down an analogous formula for a logistic regression function $\tilde{p}(x, a, b)$ with 3D input x. That is, find a formula for a logistic function

$$p : \mathbb{R}^3 \times \mathbb{R}^3 \times R \longrightarrow \mathbb{R}$$

that maps the input $x \in \mathbb{R}^3$ and parameters $a \in \mathbb{R}^3$ and $b \in \mathbb{R}$ to a probability.

b. Generalize this by writing a formula for a logistic function with an n-dimensional input x:

$$p : \mathbb{R}^n \times \mathbb{R}^n \times R \longrightarrow \mathbb{R}.$$

Exercise 3 (Monotonicity). Explain why logistic regression does not work well when $p(x)$ is not monotone.

Exercise 4 (Overfitting in linear regression). Consider data points $\{(x_1, y_1), \ldots, (x_N, y_N)\}$ with $x_i \in R^n$ and $y_i \in \mathbb{R}^m$ to which we apply linear regression; that is, we are looking for a matrix $A \in M_{m,n}(\mathbb{R})$ and a vector $b \in \mathbb{R}^m$ that minimize the least squares

$$\sum_{i=1}^{m} \|y_i - A x_i - b\|^2,$$

where $\|.\|$ denotes the Euclidian norm on $\mathbb{R}^m$.

a. Find the explicit linear equations that the matrix A and the vector b need to solve.

b. From now on, assume that $m = 1$, that is, y_i are scalar in $\mathbb{R}$ and the matrix A becomes a vector $a \in \mathbb{R}^n$. We assume moreover that $n \geq N$, so that we have more variables than data points, that is, the dimensionality n of each x_i is greater than the number N of data points, and that $(x_1, \ldots, x_N)$ is a free family in $\mathbb{R}^n$. Prove that it is possible to choose $a \in \mathbb{R}^n$ and $b \in \mathbb{R}$ such that we have exactly $y_i = a \cdot x_i + b$ for all i.

c. (robustness with respect to measurements). In part b, we saw that if $n \geq N$, then we can find a straight line which passes through all data points. Suppose that these data come from experimental measurements. Assume that the true value y_1 was measured erroneously as $\tilde{y}_1$ with small error $|\tilde{y}_1 - y_1| < \varepsilon$. Consider two lines: the unknown true line $y = ax + b$ passing through (x_1, y_1) and the line $y = \tilde{a}x + \tilde{b}$ passing through $(x_1, \tilde{y}_1)$. Would these two lines be close? That is, will $|\tilde{a} - a|$ and $|\tilde{b} - b|$ be small?

Exercise 5 (Overfitting in 1D polynomial regression). Given data points $(x_1, y_1), \ldots, (x_n, y_n)$, where $x_i, y_i \in \mathbb{R}$, $i = 1, \ldots, n$, it is always possible to find a polynomial $f(x)$ of degree $(n - 1)$ such that $f(x_i) = y_i$ for each $i = 1, \ldots, n$ (as long as $x_i \neq x_j$ whenever $i \neq j$) and of course,

$$\sum_{i=1}^{n} |f(x_i) - y_i|^2 = 0.$$

a. Find the coefficients a, b, c of the parabola $y = ax^2 + bx + c$ passing through the points $(-1, 4)$, $(0, 2)$, and $(1, -8)$.

b. Suppose $(x_1, y_1), \ldots, (x_n, y_n)$ is a large dataset (that is, $n \gg 1$) and $f(x)$ is an $(n - 1)$-degree polynomial such that $f(x_i) = y_i$ for each $i = 1, \ldots, n$.
We now continue measurements to obtain new data points. Would the polynomial fit the new data well? Explain why or why not. You can answer by way of a simple example such as fitting four data points in the plane with a cubic polynomial versus fitting with a straight line via linear regression.

Hint: Choose three data points in a straight line and one more data point near the middle one, not on the line but close. What does the cubic (third-degree polynomial) regression look like? You can use software or an online tool to perform numerical experiments, but specify what tool you used.

Exercise 6 (Understanding cross-entropy). The purpose of this exercise is to illustrate why cross-entropy plays an important role in problems such as logistic regression. For this, we revisit one example given in class about groups of students studying for an exam.

We are given N data points (x_i, r_i) for each student $i = 1 \dots N$, where x_i is the number of hours the corresponding student has spent studying and $r_i = 0$ if they failed or $r_i = 1$ if they passed. We sort the data points such that $r_i = 1$ for $i = 1, \dots, k < N$ and $r_i = 0$ for $i = k + 1, \dots, N$.

Choose any parameterization function $\bar{p}(x, \theta)$ for some parameter θ. We first consider the general optimization problem that consists in minimizing

$$L(\theta) = -\sum_{i=1}^{k} \log \bar{p}(x_i, \theta) - \sum_{i=k+1}^{N} \log(1 - \bar{p}(x_i, \theta)).$$

a. We reformulate by considering for a given number of hours x the number of students $n(x)$ who studied x hours and passed and the number of students $f(x)$ who studied x hours and failed. We also denote by $m(x) = n(x) + f(x)$ the total number of students who studied x hours. Show that we also have

$$L(\theta) = -\sum_{x=0}^{\infty} n(x) \log \bar{p}(x, \theta) - \sum_{x=0}^{\infty} f(x) \log(1 - \bar{p}(x, \theta)). \tag{6.9}$$

b. Denote $p(x) = n(x)/m(x)$. Prove that one may deduce from (6.9) that

$$\begin{aligned} L(\theta) = \bar{L} &+ \sum_{x=0}^{\infty} m(x) \left(p(x) \log \frac{p(x)}{\bar{p}(x, \theta)} + p(x, \theta) - p(x) \right) \\ &+ \sum_{x=0}^{\infty} m(x) \left((1 - p(x)) \log \frac{1 - p(x)}{1 - \bar{p}(x, \theta)} + (1 - p(x, \theta)) - (1 - p(x)) \right), \end{aligned} \tag{6.10}$$

where

$$\bar{L} = -\sum_{x=0}^{\infty} m(x) \, p(x) \log p(x) - \sum_{x=0}^{\infty} m(x) \, (1 - p(x)) \log(1 - p(x)).$$

c. Prove that for any $a, b > 0$, we have

$$a \log \frac{a}{b} + b - a \geq 0.$$

Conclude from (6.10) that $L(\theta) \geq \bar{L}$ and $L(\theta) = \bar{L}$ if and only if $p(x, \theta) = p(x)$ for all $x \geq 0$, where $m(x) > 0$.

d. Unfortunately, it may not be always possible to find a parameter θ such that $p(x, \theta) = p(x)$ for all $x \geq 0$. From now on, we specifically focus on the case of logistic regression where $\theta = (a, b)$ and

$$\bar{p}(x,\theta) = \frac{1}{1 + e^{-(ax_i+b)}}.$$

Show that if $\theta = (a, b)$ minimizes $L(\theta)$ as given by (6.9), then we have

$$\begin{aligned} 0 &= -\sum_{x=0}^{\infty} n(x)\,x\,\frac{e^{-(ax+b)}}{1+e^{-(ax+b)}} + \sum_{x=0}^{\infty} f(x)\,x\,\frac{1}{1+e^{-(ax+b)}}, \\ 0 &= -\sum_{x=0}^{\infty} n(x)\,\frac{e^{-(ax+b)}}{1+e^{-(ax+b)}} + \sum_{x=0}^{\infty} f(x)\,\frac{1}{1+e^{-(ax+b)}}. \end{aligned} \tag{6.11}$$

e. Rewrite the equations in (6.11) as

$$\begin{aligned} \sum_{x=0}^{\infty} n(x)\,x &= \sum_{x=0}^{\infty} m(x)\,x\,\frac{1}{1+e^{-(ax_i+b)}}, \\ k &= \sum_{x=0}^{\infty} m(x)\,x\,\frac{1}{1+e^{-(ax_i+b)}}. \end{aligned}$$

f. Deduce that if (a_1, b_1) and (a_2, b_2) both minimize $L(a, b)$, then we cannot have $a_1 < a_2$ and $b_1 < b_2$.

7 Support vector machine

7.1 Preliminaries: convex sets and their separation, geometric Hahn–Banach theorem

A subset A of $\mathbb{R}^n$ is called *convex* if for two points $a, b \in A$, the subset A contains the whole line segment that joins them. For example, an interval (a, b) is convex in $\mathbb{R}$, squares and disks are convex in $\mathbb{R}^2$, and cubes and balls are convex in $\mathbb{R}^3$, see Fig. 7.1.

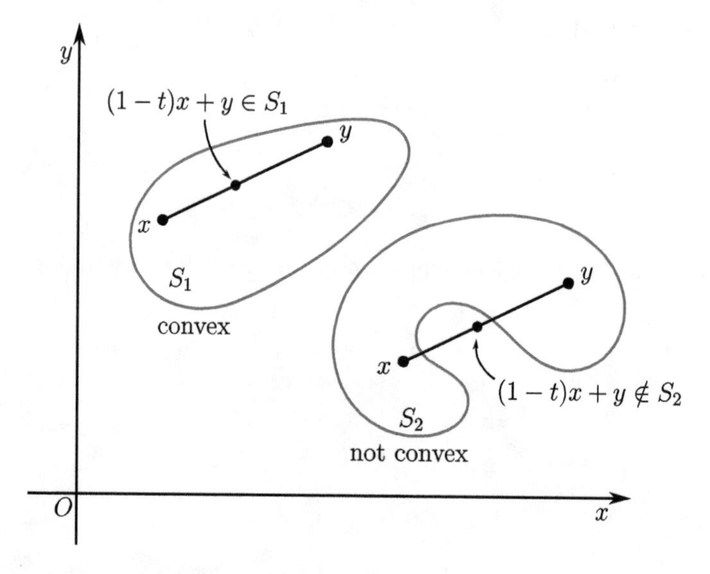

Figure 7.1: Convex and non-convex sets in $\mathbb{R}^2$.

Definition 7.1.1. $A \subset \mathbb{R}^n$ is convex if for all $x, y \in A$ the linear combination $(1-t)x + ty$ is also in A for all $t \in [0, 1]$.

An *affine hyperplane* is a subset H of $\mathbb{R}^n$ of the form

$$H = \{\boldsymbol{x} \in \mathbb{R}^n \mid \boldsymbol{w} \cdot \boldsymbol{x} - b = 0\}, \tag{7.1}$$

where $\boldsymbol{w} \in \mathbb{R}^n$ $(\neq \boldsymbol{0})$ and $b \in \mathbb{R}$. For example in $\mathbb{R}^3$, letting $\boldsymbol{w} = (w_1, w_2, w_3)$, H is the plane consisting of all points $\boldsymbol{x} = (x, y, z)$ satisfying

$$\boldsymbol{w} \cdot \boldsymbol{x} - b = w_1 x + w_2 y + w_3 z - b = 0. \tag{7.2}$$

Let A and B be two subsets of $\mathbb{R}^n$. We say that the hyperplane $H = \{\boldsymbol{x} \in \mathbb{R}^n \mid \boldsymbol{w} \cdot \boldsymbol{x} - b = 0\}$ *separates* A and B if

$$\boldsymbol{w} \cdot \boldsymbol{x} - b \leq 0 \quad \forall \boldsymbol{x} \in A \tag{7.3}$$

https://doi.org/10.1515/9783111025551-007

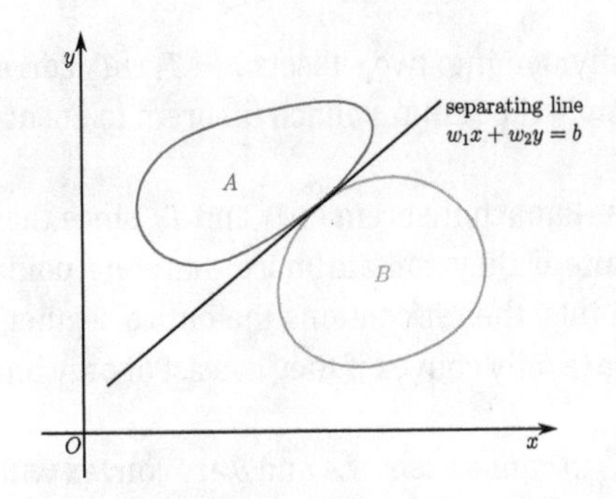

(a) Non-strict separation

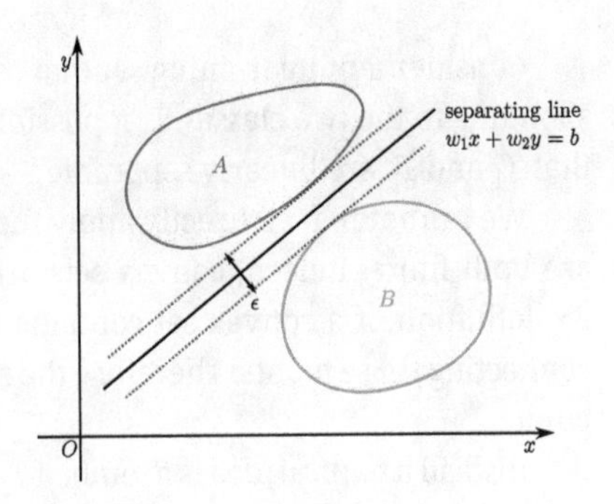

(b) Strict separation

Figure 7.2: Non-strict and strict separation of two convex sets.

and

$$\mathbf{w} \cdot \mathbf{x} - b \geq 0 \quad \forall \mathbf{x} \in B. \tag{7.4}$$

We say that H strictly separates A and B if there exists some $\varepsilon > 0$ such that

$$\mathbf{w} \cdot \mathbf{x} - b \leq -\varepsilon \quad \forall \mathbf{x} \in A \tag{7.5}$$

and

$$\mathbf{w} \cdot \mathbf{x} - b \geq \varepsilon \quad \forall \mathbf{x} \in B. \tag{7.6}$$

Geometrically, the separation means that A lies on one side of the hyperplane and B lies on the other. For example, see Fig. 7.2, where a line $w_1x + w_2y - b = 0$ separates two sets. Strict separation means that the sets lie not only on different sides of the hyperplane, but in addition, they are separated by a strip of width 2ε centered on this hyperplane. We will often refer to ε as the margin in such a case.

When A and B are convex, the famous Hahn–Banach theorem guarantees that it is possible to separate the sets.

Theorem 7.1.1 (Hahn–Banach, first geometric form: separation). *Let $A \subset \mathbb{R}^n$ and $B \subset \mathbb{R}^n$ be two non-empty convex subsets such that A and B do not intersect, that is, $A \cap B = \varnothing$. Assume that one of them is open. Then there exists a hyperplane $H = \{\mathbf{x} \in \mathbb{R}^n \mid \mathbf{w} \cdot \mathbf{x} - b = 0\}$ that separates A and B.*

The proof of the theorem can be found in [5]. We also state the version of the theorem ensuring strict separation.

Theorem 7.1.2 (Hahn–Banach, second geometric form: strict separation). *Let $A \subset \mathbb{R}^n$ and $B \subset \mathbb{R}^n$ be two non-empty convex subsets such that $A \cap B = \varnothing$. Assume that A is closed and B is closed and bounded (compact). Then there exists a closed hyperplane $H = \{\mathbf{x} \in \mathbb{R}^n \mid \mathbf{w} \cdot \mathbf{x} - b = 0\}$ that strictly separates A and B.*

Consider a finite training set that can be divided into two subsets $T = T_1 \cup T_2$ corresponding to the two classes. Is it possible to apply the Hahn–Banach theorem to ensure that T_1 and T_2 are linearly separable?

We cannot in fact directly apply the Hahn–Banach theorem to T_1 and T_2, since they are both finite. Indeed, convex sets are infinite if they contain more than one point. By definition, if a convex set contains two points, then it contains the entire segment connecting these points. Therefore the sets T_i are only convex if they consist of only one point.

Instead a natural idea is to embed T_1 and T_2 in convex sets. If A and B are convex with empty intersection, $T_1 \subset A$, and $T_2 \subset B$, then the Hahn–Banach theorem can provide a decision boundary that separates A and B and hence T_1 and T_2. We can expect that in this case, a machine learning algorithm will be able to find a classifier which separates T_1 and T_2 with a linear decision boundary.

To make this idea precise, we introduce the notion of *convex hull.*

Definition 7.1.2. Let $C \subset \mathbb{R}^n$ and let $\mathcal{A}$ be the collection of all convex subsets of $\mathbb{R}^n$ that contain C. Then the *convex hull* $\hat{C}$ of C is the intersection of all the sets in $\mathcal{A}$:

$$\hat{C} = \bigcap_{A \in \mathcal{A}} A. \tag{7.7}$$

The convex hull of C is the smallest convex subset of $\mathbb{R}^n$ that contains C. That is, every convex subset of $\mathbb{R}^n$ that contains C must also contain $\hat{C}$. For a 2D bounded subset of the plane, the convex hull may be visualized as the shape enclosed by a rubber band stretched around this set.

While T_1 and T_2 are not convex, their convex hulls $\hat{T}_1$ and $\hat{T}_2$ are. However, while T_1 and T_2 do not intersect, $\hat{T}_1$ and $\hat{T}_2$ might intersect (imagine, for instance, the convex hulls of the blue and red points in Fig. 6.12). Recall that two convex sets not intersecting is a requirement of Theorems 7.1.1–7.1.2. Therefore, by the Hahn–Banach theorem, a machine learning algorithm is able to find a linear decision boundary separating T_1 and T_2 if and only if their convex hulls $\hat{T}_1$ and $\hat{T}_2$ are disjoint. For an in-depth presentation of these and related concepts in functional analysis, the reader is referred to, for example, [5, 25, 44].

7.2 Support vector machine

Support vector machines (SVMs) are straightforward learning algorithms which can be used to separate objects (data points) into classes. In their most basic form, two classes are separated by finding a separating line in $\mathbb{R}^2$ such that each class falls onto different sides of the line (or in $\mathbb{R}^n$, each class is on different sides of the hyperplane). SVMs are examples of early machine learning algorithms that can be viewed as a precursor to more sophisticated machine learning algorithms such as DNNs.

The first instance of an SVM algorithm can be traced back to the 1963 paper by Vapnik [52] in which this algorithm was introduced and used to separate two classes of objects by a hyperplane. Later on, a detailed proof of this result was presented in the paper by Vapnik and Chervonenkis [53]. It is important to note that SVMs have been generalized to higher dimensions and using non-linear separating surfaces instead of hyperplanes (see for instance the 1992 paper by Vapnik, Boser, et al. [3]). A comprehensive introduction to SVM and related kernel methods can be found in [45].

In these notes, we will consider the simplest case of two classes in the context of supervised learning. For example, consider a set S of objects in $\mathbb{R}^2$ such that each $s \in S$ belongs to one of two classes, e. g., classes "red" and "blue"; see Fig. 7.3a. Similar to logistic regression (e. g., Fig. 6.8), an SVM will try to separate the classes through hyperplanes (e. g., straight lines in 2D). However SVMs can be seen as hard classifiers, that is, each object maps to a class. In contrast, a logistic model can be interpreted as a soft classifier because it assigns each object the probability that the object belongs to a class.

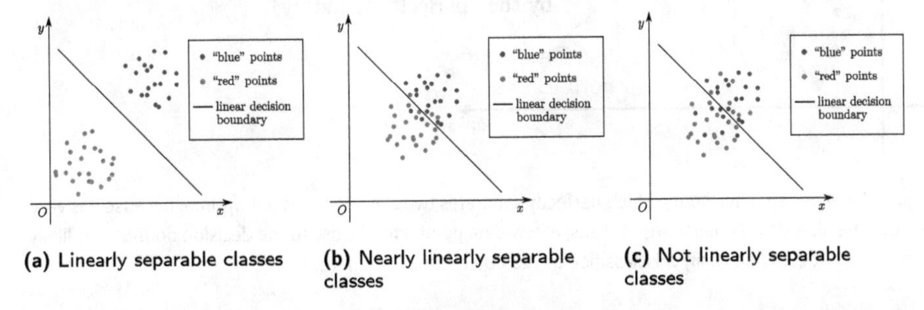

(a) Linearly separable classes **(b)** Nearly linearly separable classes **(c)** Not linearly separable classes

Figure 7.3: Linear decision boundaries in three binary classification problems. In each problem, a classifier attempts to classify all points on one side of the decision boundary as "red" and all points on the other side as "blue." In the first case, it perfectly classifies all the given points. In the second case, it cannot perfectly classify all points, but can approximately do so. In the third case, there is much more overlap between the two classes and the decision boundary can barely classify half the given points correctly.

We can distinguish two types of SVM classifier: hard margin and soft margin. A *hard margin SVM classifier* finds a linear decision boundary which perfectly separates points into their two respective classes. This is possible when the two classes are linearly separable, as in Fig. 7.3a. When the two classes are not linearly separable, as in Figs. 7.3b and 7.3c, one may use a *soft margin SVM classifier*, which identifies a linear decision boundary which approximately separates the classes (see also the discussion on the decision boundary in Section 6.3.2). In that sense, a soft margin SVM classifier is somewhat similar to a logistic model (cf. Fig. 6.8) trained with a different loss function described below (cf. cross-entropy loss (6.7) in regression).

By contrast, a hard margin SVM classifier further differs from logistic models in that it identifies the linear decision boundary which separates the two classes by the *widest margin*, which is a rather different optimization process. To ensure such a widest possible margin, the decision boundary of a hard margin SVM classifier is chosen in such a way that the nearest point of each class is as far away as possible from the decision boundary; see Fig. 7.6. Having a wide margin is obviously beneficial, as illustrated in

Fig. 7.4, where a decision boundary that is too close to one of the two classes is for example more likely to misclassify new data points whose class is unknown.

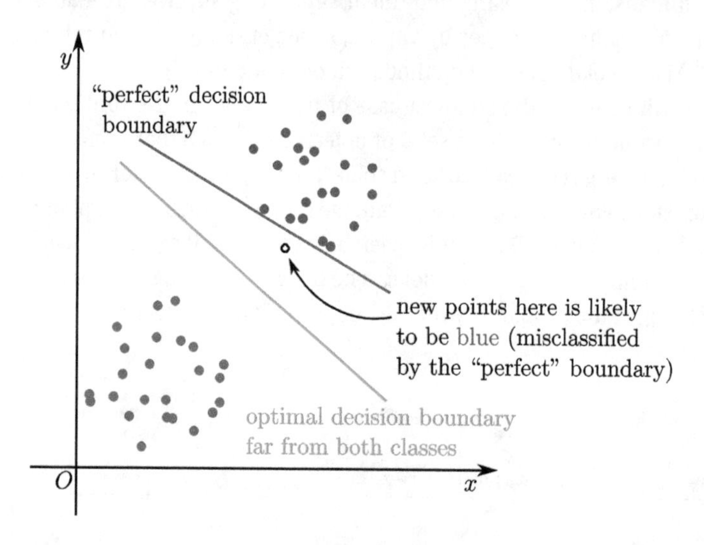

Figure 7.4: A decision boundary which perfectly separates two classes, but is not optimal because it is very close to the blue class (small margin). Thus, a new data point added close to the decision boundary is likely to be blue, but could be easily misclassified as red due to the small margin.

In a classification problem where the two classes can be perfectly separated by a linear decision boundary on a training set, a logistic regression algorithm may minimize the cross-entropy loss by making a sharp transition of $\tilde{p}(x, y)$ from 0 to 1. In such a situation the choice of the decision boundary has much less impact on the loss function than the sharpness of the transition; see Fig. 7.4. Thus, in such problems, SVM is a better choice because it will focus on identifying the optimal decision boundary and be more likely to correctly classify new objects on the testing set.

7.3 Hard margin SVM classifiers

As mentioned above, we use hard margin SVM classifiers when one has two classes of objects which can be separated by a linear decision boundary, as in Fig. 7.3a; in such a case we say that the classes are *linearly separable*. But even in cases of linearly separable classes, there are of course many lines that separate the classes; see Fig. 7.5. Thus, the goal of a hard margin SVM classifier is to find the optimal linear decision boundary in some appropriate sense.

We start with a training set $(x_1, y_1), \ldots, (x_N, y_N)$, with $x_i \in \mathbb{R}^n$ denoting the objects and $y_i \in \{-1, 1\}$ representing the class to which x_i belongs (e. g., 1 for "red" and -1 for "blue" in Fig. 7.3a). For simplicity we will assume that x_i are points in $\mathbb{R}^2$. The goal of hard margin SVM classifiers is twofold:

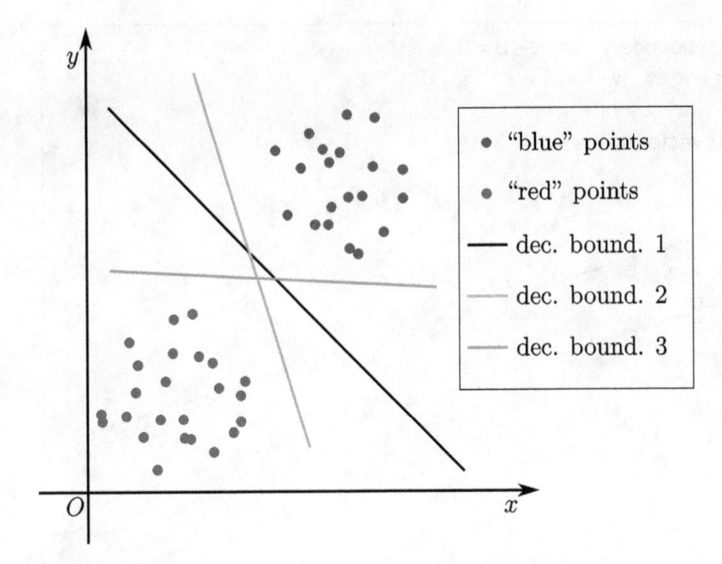

Figure 7.5: When two classes can be separated by a linear decision boundary, there are typically infinitely many such decision boundaries, leading to the question which is best.

1. separate the classes by a linear decision boundary of the form $\mathbf{w} \cdot \mathbf{x} - b = 0$, where $\mathbf{w} \in \mathbb{R}^2$ and $b \in \mathbb{R}$;
2. find the parameters w and b such that the linear decision boundary has the largest possible *margin* which excludes all x_i.

The margin is the set of all points in $\mathbb{R}^2$ between two parallel lines defined as follows:

$$\{\mathbf{x} \in \mathbb{R}^2 : -1 < \mathbf{w} \cdot \mathbf{x} - b < 1\}; \tag{7.8}$$

see Fig. 7.6 (cf. strict separation of convex sets in Section 7.1). The distance between these two lines is $\frac{2}{\|w\|}$. Therefore, finding the largest margin that excludes all x_i means minimizing $\|w\|$ while ensuring that $\mathbf{w} \cdot \mathbf{x}_i - b \geq 1$ if $y_i = 1$ and $\mathbf{w} \cdot \mathbf{x}_i - b \leq -1$ if $y_i = -1$ for each $i = 1, \ldots, N$. Thus the optimal w and b are found by solving the following constrained optimization problem ($y_i = \pm 1$ for red and blue):

$$\min_{\mathbf{w} \in \mathbb{R}^2, b \in \mathbb{R}} \|w\| \quad \text{such that} \quad y_i(\mathbf{w} \cdot \mathbf{x_i} - b) \geq 1 \quad \forall i = 1, \ldots, N. \tag{7.9}$$

If w and b solve (7.9), then the decision boundary $\mathbf{w} \cdot \mathbf{x} - b = 0$ is optimal in the following sense. First, all data points x_i such that $y_i = 1$ are on one side of the line while all x_i such that $y_i = -1$ are on the other side of the line (i. e., the line perfectly separates the two classes). Second, among all linear decision boundaries which perfectly separate both classes (of which there are many; see Fig. 7.5), this one has the maximum distance from the nearest elements $\mathbf{x}_i$ from each class. This is because all objects $\mathbf{x}_i$ are outside the

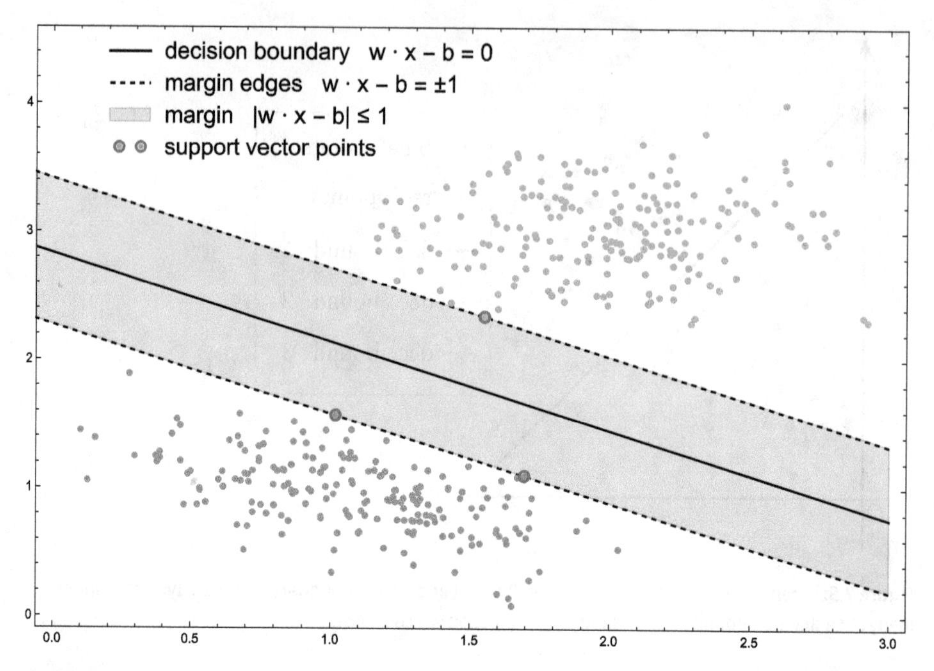

Figure 7.6: An example of classification using SVM. The red circled dots are the support vectors.

margin. The distance from the decision boundary to the margin is $1/\|\boldsymbol{w}\|$, so by minimizing $\|\boldsymbol{w}\|$ in (7.9), we maximize the distance from the decision boundary to the objects $\boldsymbol{x}_i$.

These nearest elements are called the *support vectors*, which are illustrated in Fig. 7.6. Observe that the support vectors must all lie on the boundary of the margin; otherwise the margin can be made wider while still satisfying the restriction given in (7.9). The name “support vector” comes from the fact that the support vectors alone determine the optimal SVM decision boundary in the following sense. If any object x_i which is not a support vector is removed from the dataset $\{x_1, \ldots, x_N\}$, then the optimal SVM decision boundary does not change. That is, we may remove all constraints in (7.9) except those corresponding to the support vectors:

$$\min_{\mathbf{w} \in \mathbb{R}^2, b \in \mathbb{R}} \|w\| \quad \text{such that} \quad y_i(\mathbf{w} \cdot \mathbf{z_i} - b) = 1 \quad \forall i = 1, \ldots, k, \tag{7.10}$$

where $\mathbf{z}_1, \ldots, \mathbf{z}_k \in \{\boldsymbol{x}_1, \ldots, \boldsymbol{x}_N\}$ are the support vectors.

7.4 Soft margin SVM classifier

In the somewhat typical case where no linear decision boundary exactly separates two classes, as in Fig. 7.3b, a soft margin SVM classifier can still be used to find a linear decision boundary that best separates the two classes, even if it is not perfect.

Soft margin SVM classifiers make use of the *hinge loss function* for individual objects $\mathbf{x}_i$:

$$L_i(\mathbf{w}, b) = \mathrm{ReLU}(1 - y_i(\mathbf{w} \cdot \mathbf{x}_i - b)), \quad i = 1, \ldots, N. \tag{7.11}$$

In the soft margin SVM classifier, we do not require the constraint in (7.9) to hold. But if $y_i(\mathbf{w} \cdot \mathbf{x}_i - b) \geq 1$ for some i, $\mathbf{w}$, b, that is, if x_i does satisfy the constraint, then $L_i(\mathbf{w}, b) = 0$ because $\mathrm{ReLU}(x) = 0$ when $x < 0$. Conversely, if $y_i(\mathbf{w} \cdot \mathbf{x}_i - b) < 1$, then $L_i(\mathbf{w}, b) > 0$, increasing linearly with $1 - y_i(\mathbf{w} \cdot \mathbf{x}_i - b)$. Our goal is to find a linear decision boundary such that:

- Most $\mathbf{x}_i$ lie on the correct side of the decision boundary by a distance of at least $1/\|\mathbf{w}\|$ (i. e., $L_i(\mathbf{w}, b) = 0$).
- Some $\mathbf{x}_i$ will be closer to the decision boundary than $1/\|\mathbf{w}\|$, but still on the correct side (i. e., $0 < L_i(\mathbf{w}, b) \leq 1$).
- Even fewer $\mathbf{x}_i$ may even be on the wrong side of the decision boundary (i. e., $L_i(\mathbf{w}, b) > 1$), but not many.

To try to ensure this behavior, we introduce the regularized average hinge loss function:

$$L(\mathbf{w}, b) = \lambda \|\mathbf{w}\|^2 + \frac{1}{N} \sum_{i=1}^{N} L_i(\mathbf{w}, b), \tag{7.12}$$

with an additional term $\lambda \|\mathbf{w}\|^2$.

To understand the meaning of (7.12), it is useful to consider how a soft margin SVM classifier will be used. One typically assigns to the corresponding class any object that is a distance of at least $1/\|\mathbf{w}\|$ away from the decision boundary. Objects that lie closer to the decision boundary (inside the margin) are not considered to be reliably classified.

Therefore a soft margin SVM classifier tries to satisfy two somewhat competing objectives. First of all, it should be reliable: most objects that are assigned to a class should indeed belong to that class. Second, it should be reasonably precise: we want to have as few objects as possible that are not reliably classified.

Consider now the cases $\lambda \ll 1$ and $\lambda \gg 1$ in (7.12). If λ is large, minimizing (7.12) mostly consists of trying to take $\|w\|$ as small as possible, i. e., trying to have a large margin. However, we then give little weight to the terms involving $L_i(\mathbf{w}, b)$. For this reason, we may have a relatively large number of objects that are not reliably classified. On the other hand, because the margin is quite large, we likely do not have many objects that are wrongly classified. Such a classifier may be reliable but not precise at all.

On the other hand, if λ is very small, we focus on minimizing $L_i(\mathbf{w}, b)$ on the training set. On the training set, this optimization will try to limit the number of misclassified objects but also the number of objects within the margin. Hence it may lead to a very small margin. When looking at new objects outside of the training set, this likely leads to a high error rate: An object that is at a small distance from a well-classified object in the training set in the same class may end on the wrong side of the decision boundary if the margin is even smaller. We would obtain a precise classifier that is not reliable.

Therefore, one must find an appropriate value of λ (not too big or too small) in a trial-and-error process often referred to as "tuning." The parameter λ is an example of a *hyperparameter*. In machine learning, a hyperparameter is a value which controls the learning process, as opposed to a value determined by the learning process. This can be compared to hyperparameters in DNNs such as the number of layers and their width which control the learning process versus the parameters of a DNN obtained by training.

Example 7.4.1. Graduate school admission example revisited with SVM.

We now revisit the graduate school admission example from Section 6.3.2 to better illustrate the similarities and differences between logistic models and SVMs. Recall that we have a dataset consisting of pairs of GPAs and GRE scores for many graduate school applicants, along with data on whether each applicant was accepted or not. We wish to find a good model to try to predict which students will be accepted. In Section 6.3.2, we used logistic regression; here we present the results for a soft margin SVM classifier.

To this end we minimize (7.12) over all $\boldsymbol{w} \in \mathbb{R}^2$ and $b \in \mathbb{R}$, choosing λ (after some trial-and-error) to be 0.005. Just as for the logistic model, a key feature of the resulting classifier is the decision boundary shown in Fig. 7.7. The decision boundary found by the soft margin SVM classifier is very similar to that found by logistic regression in Fig. 6.8.

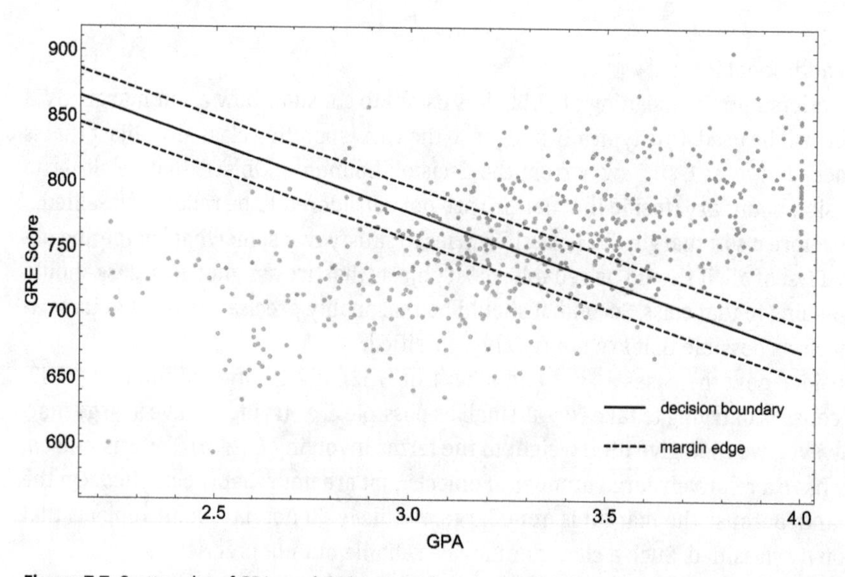

Figure 7.7: Scatter plot of GPAs and GRE scores of graduate school applicants. Blue points represent applicants who were ultimately accepted, while orange points represent applicants who were rejected. The solid black line is the decision boundary identified by a soft margin SVM classifier, and the dashed lines bound the associated margin.

Recall from Section 6.3.2 that the logistic regression model provides a predicted probability function for this problem. In contrast, the soft margin SVM classifier is closer to a hard classifier since it maps objects (e. g., the pairs (GPA, GRE)) to classes (e. g., accepted or not accepted) rather than to probabilities. But, in addition to identifying the classes, the soft margin SVM classifier does provide a margin, as can be seen in Fig. 7.7. This still offers a natural interpretation: Applicants who are above the margin are predicted to be accepted, and applicants who are below the margin are predicted to be declined. The SVM classifier typically does not attempt to make a prediction for applicants within the margin; indeed, as can be seen, the two classes overlap to a considerable extent in the margin.

7.5 Exercises

Exercise 1 (A basic example of separation). Consider the equation for the line (1D hyperplane)

$$w \cdot x - b = 0, \tag{7.13}$$

where $w = (1, -1)$ and $b = 1$.

a. Draw a graph of the line (7.13).
b. Explain if the line (7.13) separates the following two circular disks, A and B:

$$A = \{(x, y) \in \mathbb{R}^2 \,:\, (x-1)^2 + (y-1)^2 < 1/2\},$$
$$B = \{(x, y) \in \mathbb{R}^2 \,:\, (x-2)^2 + y^2 < 1/2\}.$$

Exercise 2 (An example of hard margin SVM). Consider the following data points: $(x_1 = (-1, 0), y_1 = -1)$, $(x_2 = (1, 0), y_1 = 1)$, $(x_3 = 0, y_3 = -1)$, $(x_4 = (2, 0), y_4 = 1)$, where x_i belong to $\mathbb{R}^2$ and y_i belong to $\mathbb{R}$.

a. Construct the hard margin classifier corresponding to the two classes labeled $y_i = -1$ or $y_i = 1$. Give the support vector.
b. We add a fifth data point $(x_5 = (-1, 1), y_5 = -1)$. Explain why the hard margin classifier is identical to the previous question.
c. We add a sixth data point $(x_5 = (1/2, 1), y_5 = 1)$. Explain why the hard margin classifier now has to be different and calculate it.

Exercise 3 (Existence and non-existence of hard margin classifiers). Consider the following partial data points: $x_1 = (0, 0)$, $x_2 = (1, 0)$, $x_3 = (1, 1)$, $x_4 = (2, 1)$.

a. We also consider two classes labeled $y_i = -1$ or $y_i = 1$ with $y_1 = 1$, $y_2 = -1$, $y_3 = -1$, $y_4 = 1$.
   a.1. Show that there does not exist a hard margin classifier.
   a.2. Determine the convex hull of each class and their intersection.
b. We now take $y_1 = 1$, $y_2 = 1$, $y_3 = -1$, $y_4 = -1$.
   b.1. Show that there exists a hard margin classifier.

b.2. Determine the convex hull of each class and their intersection.

c. Study all possible choices for y_i and determine for which a hard margin classifier exists and for which it does not. Compare this with what the intersection of the convex hull of each class predicts.

Exercise 4 (An example of soft margin classifier). Consider the following data points: $(x_1 = (-1, 0), y_1 = -1)$, $(x_2 = (1, 0), y_1 = 1)$, $(x_3 = 0, y_3 = -1)$, $(x_4 = (2, 0), y_4 = 1)$, where x_i belong to $\mathbb{R}^2$ and y_i are either -1 or 1 corresponding to two separate classes. We wish to find the soft margin classifier corresponding to this example.

a. Write the average hinge loss function and the regularized average hinge loss function $L(w, b)$ that we have to minimize.

b. Recall that the ReLU function is smooth outside of 0 and calculate its derivative.

c. Identify the subdomains where $L(w, b)$ is smooth.

d. Calculate the gradient of $L(w, b)$ on each such subdomain.

e. Identify the minimum of $L(w, b)$.

Exercise 5 (Hard vs. soft margin classifier). We consider $N = 2n$ data points where $x_1 = \cdots = x_n = (0, 0)$ with $y_1 = \cdots = y_n = -1$ and $x_{n+1} = \cdots = x_{2n} = (1, 0)$ with $y_{n+1} = \cdots = y_{2n} = 1$.

a. Calculate the hard margin classifier for this dataset.

b. Calculate the soft margin classifier for this dataset.

c. We now add one point to the dataset: $x_{2n+1} = (1/2, 0)$ with $y_{2n+1} = -1$. Calculate the new hard margin classifier.

d. Show that for n large, the new soft margin classifier with the additional (x_{2n+1}, y_{2n+1}) remains close to the classifier identified in question b.

Exercise 6 (Support vectors for hard margin SVMs). We consider data points x_i for $i = 1 \ldots N$ in $\mathbb{R}^2$ belonging to two classes labeled $y_i = -1$ or $y_i = 1$. We assume that there exists at least one optimal hard margin classifier given by $\bar{w}$ and $\bar{b}$.

a. Assume that three support vectors exist, two from class -1 corresponding to points x_1 and x_2 and one from class $+1$ corresponding to point x_3.

a.1. Prove that the direction $\bar{w}/\|\bar{w}\|$ is uniquely determined by x_1 and x_2.

a.2. Prove that both $\bar{w}$ and $\bar{b}$ are uniquely determined by x_1, x_2, and x_3.

b. We now assume that only two support vectors exist, x_1 for class -1 and x_2 for class $+1$.

b.1. Show that there exists an infinite set S of possible (w, b) such that $w \cdot x_1 - b = -1$ and $w \cdot x_2 - b = 1$.

b.2. Explain that among all (w, b) in S the element (w_0, b_0) with the minimum norm $\|w_0\|$ is the one such that w_0 is parallel to the line connecting x_1 and x_2.

b.3. Show that there exists a small but still infinite set $S' \subset S$ that perfectly separates the two classes and uses x_1 and x_2 as support vectors.

b.4. Out of all the possible choices of (w, b) in S', explain why there is only one that minimizes $\|w\|$ and why this is necessarily the optimal hard margin classifier $(\bar{w}, \bar{b})$.

b.5. Show why we need to have $w_0 = \bar{w}$ if the hard classifier has no other support vectors outside of x_1 and x_2.

c. Explain why there cannot be only one support vector. Give an example with four support vectors.

8 Gradient descent method in the training of DNNs

In several regards, training is the key feature of machine learning algorithms since this is when the algorithms learn the data; see Chapters 3–7. In the context of *supervised learning*, the training of DNNs can often be represented through an optimization process that consists of the minimization of a so-called loss function $L(\boldsymbol{\alpha})$. Here $\boldsymbol{\alpha}$ is the vector of DNN parameters that was introduced in Definition 4.1.3 within the simple framework that we consider.

Critical or stationary points of a multivariable function $L(\boldsymbol{\alpha})$ are the points $\boldsymbol{\alpha}$ solving the equation $\nabla L(\boldsymbol{\alpha}) = 0$. Critical points of course include local minima, local maxima, and saddle points, which can be distinguished by the second derivative test (testing the positive definiteness of the Hessian matrix).

For simple functions in low dimension, it is sometimes possible to use this to analytically find local minima or even global minima by comparing the values of $L(\boldsymbol{\alpha})$ at those local minima. For DNN loss functions, such an analytical approach is unpractical in general: The analytical expressions for the functions $L(\boldsymbol{\alpha})$, $\nabla L(\boldsymbol{\alpha})$, and $\nabla^2 L(\boldsymbol{\alpha})$ are not expected to be simple. Moreover, the dimension μ of the parameter space $\mathbb{R}^\mu$ is very large, so finding the exact solution of the system $\nabla L(\boldsymbol{\alpha}) = 0$ does not appear feasible.

Hence, in practice minimizing this type of function with a very large number of parameters is a difficult problem for which several numerical approaches have been proposed. We briefly discuss here some common numerical optimization methods for DNN training that are based on the notion of *gradient descent* (GD). The idea behind such methods is simply that a multivariable function decreases fastest in the direction of the negative gradient of the function. Heuristically, it can be compared to the steps a hiker could take to descend from a forested hillside when they cannot see the bottom of the hill through the trees. In this case, the hiker seeks to minimize their elevation while only being able to see what is nearby, that is, locally. A possible strategy, therefore, is to take each step in whichever direction leads them the fastest downhill locally without being able to take into consideration the future steps to the global minimum. As one can see from this analogy, establishing the convergence of GD methods to a global minimum could be a challenge, especially in the presence of local minima.

We will consider deterministic GD and stochastic GD (SGD) methods. We also very briefly mention here a generalization of GD called GD with *momentum* or the *heavy ball method* [41]. The idea is to add inertia to prevent "getting stuck" or "slowing down" in local minima due to oscillations in the loss function $L(\boldsymbol{\alpha})$. Both SGD and GD with momentum help reduce the computation time and help deal with local minima and saddle points.

https://doi.org/10.1515/9783111025551-008

8.1 Deterministic gradient descent for the minimization of multivariable functions

We start with the deterministic *gradient descent* (GD) method, which is an iterative process where we update the parameters at each step along the direction of the gradient of the loss function.

The GD method seems to have first been introduced by Cauchy in the mid-nineteenth century and rediscovered by Hadamard at the beginning of the twentieth century. It has since then become one of the most extensively studied numerical algorithms and has demonstrated its usefulness for finding local minima of functions; see for instance [40]. In the specific case of convex functions, several extensions have been proposed to improve the convergence rate while remaining at first order (e. g., involving only the gradient of the function to be minimized rather than higher-order derivatives), for which we refer in particular to [36].

Consider a point $\boldsymbol{a}$ together with a direction $\boldsymbol{\gamma}$. We say that a function L is decreasing in the direction of $\boldsymbol{\gamma}$ if the derivative of L in the direction of $\boldsymbol{\gamma}$ is negative, i. e.,

$$\nabla L(\boldsymbol{a}) \cdot \boldsymbol{\gamma} < 0. \tag{8.1}$$

We recall the elementary identity for the dot product on $\mathbb{R}^{\mu}$:

$$\nabla L(\boldsymbol{a}) \cdot \boldsymbol{\gamma} = \|\nabla L(\boldsymbol{a})\| \, \|\boldsymbol{\gamma}\| \cos\theta, \tag{8.2}$$

where θ is the angle between the vectors $\nabla L(\boldsymbol{a})$ and $\boldsymbol{\gamma}$. Therefore, the directional derivative (8.2) is maximized when $\theta = 0$ ($\boldsymbol{\gamma}$ is in the direction of the gradient) and minimized when $\theta = \pi$ ($\boldsymbol{\gamma}$ is in the direction of the negative gradient).

This can be used to find another point $\tilde{\boldsymbol{a}}$ from $\boldsymbol{a}$ where L is lower. By the Taylor formula, provided that L is smooth,

$$L(\boldsymbol{a} + \tau\,\boldsymbol{\gamma}) = L(\boldsymbol{a}) + \tau\,\nabla L(\boldsymbol{a}) \cdot \boldsymbol{\gamma} + O(\tau^2).$$

We give the precise definition of the big $O(\cdot)$ notation in Definition 9.1.1.

If τ is small enough, our best "guess" for a new point $\tilde{\boldsymbol{a}}$ naturally consists in taking

$$\tilde{\boldsymbol{a}} = \boldsymbol{a} + \tau\,\boldsymbol{\gamma}, \quad \text{with } \gamma = -\nabla L(\boldsymbol{a}).$$

This is the idea behind GD algorithms: From a random initial point in the parameter space, move to the next point by choosing the appropriate direction of the gradient. Continue iteratively until the desired precision in minimization is achieved.

In more mathematical terms, we begin with a randomly chosen initial parameter vector $\boldsymbol{a}^{(0)} \in \mathbb{R}^{\mu}$. Then we iteratively calculate a sequence of parameters $(\boldsymbol{a}^{(n)})_{n\in\mathbb{N}}$ according to the following rule:

$$\boldsymbol{a}^{(n)} = \boldsymbol{a}^{(n-1)} - \tau \nabla L(\boldsymbol{a}^{(n-1)}). \tag{8.3}$$

The *learning rate* τ (also called step size) governs how big the "step" from $\boldsymbol{a}^{(n-1)}$ to $\boldsymbol{a}^{(n)}$ is. It represents the speed of learning in the sense that it controls how much "new information" $\tau \nabla L(\boldsymbol{a}^{(n-1)})$ is added to the "old information" $\boldsymbol{a}^{(n-1)}$ to decide on the new parameter value $\boldsymbol{a}^{(n)}$.

The learning rate has to be small enough because the gradient $\nabla L(\boldsymbol{a})$ contains only local information about L in a small neighborhood of $\boldsymbol{a}$. In the example of the hiker, a huge miles-long "step" in the downhill direction could take the hiker to the top of another hill rather than to the bottom of their current hill. However, a learning rate that is *too* small means that we may require many more iterations than is necessary, leading to training taking an excessively long time. Therefore, it is important to strike a balance when choosing τ.

Example 8.1.1 (Choosing τ). Use GD to approximate the minimum of $F(x) = x^2$ (which is obviously attained at $x = 0$) with initial value $x_0 = 1$ and different values of τ. Table 8.1 illustrates that choosing very small τ leads to convergence taking many iterations, whereas for large τ the GD algorithm does not converge at all.

Table 8.1: Number of steps for gradient descent with different values of τ.

τ	Smallest N such that $\lvert x^n - 0\rvert < 0.001$ for all $n > N$
$\tau = 0.001$	$N = 3451$
$\tau = 0.01$	$N = 342$
$\tau = 0.1$	$N = 31$
$\tau = 0.5$	$N = 1$
$\tau = 1.0$	does not converge

Ideally the iterative steps of the GD algorithm lead to a minimizer $\bar{\boldsymbol{a}}$, a point that minimizes $L(\boldsymbol{a})$, as illustrated in Fig. 8.1. This figure also demonstrates a natural geometrical interpretation to the algorithm: At each step of the GD, the gradient is perpendicular to the nearby level set.

More specifically, if $L(\boldsymbol{a}_0) = c$, the gradient $\nabla L(\boldsymbol{a}_0)$ is perpendicular to the level set $\{\boldsymbol{a} \in \mathbb{R}^\mu : L(\boldsymbol{a}) = c\}$. This can easily be proved by considering any curve $\mathbf{r} : \mathbb{R} \to \mathbb{R}^\mu$ such that $\boldsymbol{r}(0) = \boldsymbol{a}_0$ and $L(\boldsymbol{r}(t)) = c$ for all t. Then on the one hand,

$$\frac{d}{dt} L(\boldsymbol{r}(t))\bigg|_{t=0} = \frac{d}{dt} c = 0, \tag{8.4}$$

while on the other hand,

$$\frac{d}{dt} L(\boldsymbol{r}(t))\bigg|_{t=0} = \nabla L(\boldsymbol{r}(0)) \cdot \boldsymbol{r}'(0) = \nabla L(\boldsymbol{a}) \cdot \boldsymbol{r}'(0). \tag{8.5}$$

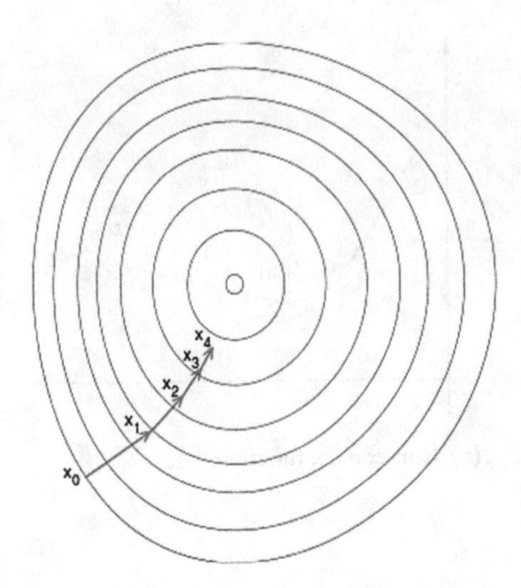

Figure 8.1: How GD works. *Blue curves:* Level set of the function $L(\boldsymbol{a}) : \mathbb{R}^2 \to \mathbb{R}$ (c-level set $= \{\boldsymbol{a} \in \mathbb{R}^2 \mid L(\boldsymbol{a}) = c\}$, c is smaller in the center). *Red arrows:* Direction of the negative gradient (perpendicular to the level set).

Since $\mathbf{r}'(0)$ can be any tangent vector to the level set $L(\boldsymbol{a}) = c$ at the point $\boldsymbol{a} = \boldsymbol{a}_0$, we conclude that $\nabla L(\boldsymbol{a}_0)$ is perpendicular to the level set.

However, a key question for GD algorithms is whether the sequence $(L(\boldsymbol{a}_n))_{n\in\mathbb{N}}$ converges to a minimal value $L(\bar{\boldsymbol{a}})$. There are in fact many reasons why convergence to a minimizer would fail: One may for example converge instead to a local minimum. But there are also cases where the algorithm completely fails to converge to any point.

Just as for separating sets, the notion of convexity is very useful to guarantee convergence when it is now applied to functions.

Definition 8.1.1. A function $L : \mathbb{R}^\mu \to \mathbb{R}$ is *convex* if

$$L(t\boldsymbol{a}_1 + (1-t)\boldsymbol{a}_2) \le tL(\boldsymbol{a}_1) + (1-t)L(\boldsymbol{a}_2), \quad \forall \boldsymbol{a}_1, \boldsymbol{a}_2 \in \mathbb{R}^\mu,\ t \in [0,1]. \tag{8.6}$$

Inequality (8.6) means that the line segment joining the points $(\boldsymbol{a}_1, L(\boldsymbol{a}_1))$ and $(\boldsymbol{a}_2, L(\boldsymbol{a}_2))$ lies above the graph of $L(\boldsymbol{a})$; see Fig. 8.2.

Convex functions commonly play an important role in optimization, not only for GD algorithms. A first helpful property is that local and global minima are identical for convex functions.

Theorem 8.1.1 (Local vs. global minimum). *Any local minimum of a convex function $L : \mathbb{R}^\mu \to \mathbb{R}$ is also a global minimum.*

The next theorem shows that GD algorithm are guaranteed to perform well on convex functions.

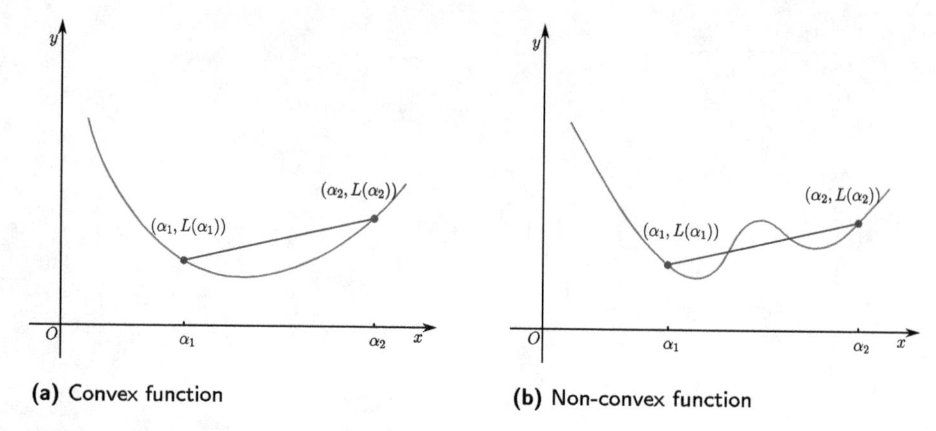

(a) Convex function

(b) Non-convex function

Figure 8.2: Convex and non-convex functions.

Theorem 8.1.2 (Convergence of GD). *Suppose the function $L : \mathbb{R}^\mu \to \mathbb{R}$ has the following properties:*

(i) *L is convex and differentiable,*

(ii) *L admits a (global) minimum at $\bar{\boldsymbol{a}} \in \mathbb{R}^\mu$, and*

(iii) *∇L is* Lipschitz *with constant d, i. e., we have $\|\nabla L(\boldsymbol{a}_1) - \nabla L(\boldsymbol{a}_2)\| \leq d\|\boldsymbol{a}_1 - \boldsymbol{a}_2\|$ for any $\boldsymbol{a}_1, \boldsymbol{a}_2$ (where $\|\cdot\|$ is the Euclidean norm).*

Assume that $\tau \leq 1/d$. Then the iterations $\boldsymbol{a}^{(k)}$ of the GD algorithm satisfy

$$L(\boldsymbol{a}^{(k)}) - L(\bar{\boldsymbol{a}}) \leq \frac{\|\boldsymbol{a}^{(0)} - \bar{\boldsymbol{a}}\|^2}{2\tau k}. \tag{8.7}$$

Intuitively, this theorem means that GD is guaranteed to converge if the learning rate is small enough, and it converges with rate $O(1/k)$.

However, there is no general rule for choosing the optimal τ. Furthermore, one can choose different values of τ_k for each step in the iteration.

8.2 Additive loss functions

The use of SGD algorithms typically requires a special structure of the loss function. We consider here so-called additive loss functions, which easily let us explain the basic SGD method. They have the form

$$L(\boldsymbol{a}) = \frac{1}{|T|} \sum_{s \in T} \ell(s, \boldsymbol{a}), \tag{8.8}$$

where $\ell(s, a)$ quantifies the quality of approximation for each individual object $s \in T$.

A simple example of function $\ell(s, \boldsymbol{a})$ is the squared distance between $\phi(s, \boldsymbol{a})$ and $\psi(s)$ in the Euclidean norm:

$$\ell(s, \boldsymbol{a}) = \|\phi(s, \boldsymbol{a}) - \psi(s)\|^2. \tag{8.9}$$

If (8.9) is used for $\ell(s, \boldsymbol{a})$ in (8.8), the loss function $L(\boldsymbol{a})$ is usually called the *mean square loss*.

In (8.8), $|T|$ is the number of objects in T, and therefore the sum $1/|T| \sum_{s \in T} \cdot$ in (8.8) represents the averaging of the loss over all elements in T. The minimization of $L(\boldsymbol{a})$ hence hopefully leads to minimization of $\ell(s, \boldsymbol{a})$ for most objects s.

For DNNs solving the classification problem, another typical choice is the *cross-entropy loss* function, which generalizes the cross-entropy loss for logistic regression (see (6.7)). Recall that if a DNN ϕ is used for classification, it has the form given in Definition 4.2.1 and $\phi(s, \boldsymbol{a}) = (p_1(s, \boldsymbol{a}), \ldots, p_m(s, \boldsymbol{a}))$, a vector of probabilities that s belongs to each of m classes. We then define the cross-entropy loss as an average over the training set T:

$$L(\boldsymbol{a}) = -\frac{1}{|T|} \sum_{s \in T} \log(p_{i(s)}(s, \boldsymbol{a})), \tag{8.10}$$

where $p_{i(s)}(s, \boldsymbol{a})$ (from (3.3) and (3.4)) is the probability predicted by the DNN that s belongs to its correct class, $i(s)$.

Each term $-\log(p_{i(s)}(s, \boldsymbol{a}))$ in the sum (8.10) is the *loss contributed by object* s. If $p_{i(s)}(s, \boldsymbol{a}) \approx 0$, then ϕ does not classify s very well and the loss contributed by s is large. On the other hand, if $p_{i(s)}(s, \boldsymbol{a}) \approx 1$, then ϕ does a good job at classifying s, and the loss contributed by s is small. By choosing $\boldsymbol{a}$ so that $L(\boldsymbol{a})$ is minimized, we expect that on average, ϕ classifies each object $s \in T$ as well as possible.

8.3 What are SGD algorithms? When to use them?

Methods based on so-called stochastic approximation have become popular in a wide range of problems, as showcased in particular in [47]. The earliest examples of such methods appear to have been introduced in [23, 42]. In the specific context of machine learning, SGD algorithms and their many variants are nowadays the standard method to optimize artificial neural networks when combined with backpropagation; see for example [15, 30]. SGD is a particularly helpful method when the training set is large, as explained for instance in [4].

Specifically, minimizing a loss function of the form (8.8) through deterministic GD involves calculating $\nabla \ell(s, \boldsymbol{a})$ for each object at each iteration step to estimate the sum over all objects. For large training sets, this necessarily results in a high computational cost.

On the other hand, the design of SGD algorithms *reduces the number of such operations*. The main idea is to choose a random sample of objects from the training set T called a *batch* to calculate the gradient at each iteration step instead of using the entire training set. Thus, while GD uses (8.3) for its iterative step, SGD uses

$$a^{(n)} = a^{(n-1)} - \tau \nabla \frac{1}{|T_n|} \sum_{s \in T_n} \ell(s, a), \tag{8.11}$$

where $T_n \subset T$ is the batch used at the nth iteration (a randomly chosen batch for each iteration).

In a simple version of SGD, the number of objects in each batch is fixed over all iterations, and this number is called the *batch size*. When compared with deterministic GD, SGD leads to the following differences:

- Computational complexity (the computational cost) is reduced since we only use some of the terms in (8.8). One can usually expect the computation time of each iteration to scale linearly with batch size.
- We do not compute the same gradient at each time step. Ideally SGD calculates a reasonable approximation of the gradient of L. But in practice, the two can be significantly different. The fewer terms we use (the smaller the batch size), the more likely this is of course. A first consequence that is often very noticeable in practice is that we may have large stochastic fluctuations for small batch sizes: From one iteration to the next, we may choose very different objects and obtain very different gradients of descent. For similar reasons, SGD algorithms may also require more time steps to converge than deterministic GD; the total computational cost may still be lower for SGD as each time step is much less costly. There are also advantages to not calculating the same gradient as SGD may be better able to avoid being trapped in local minima of the loss function.

For all these reasons, making an appropriate choice of batch size for SGD algorithms may require some trial-and-error to get a good compromise between gains in computational time and better behavior.

8.4 Epochs in SGD

The notion of epochs is motivated by the fact a batch does not "cover" the entire training set T, and is often much smaller than T. Intuitively, an epoch is the number of batches needed to cover the entire training set. More precisely, an epoch is a family of iterations of the following size:

$$\text{\# of iterations in an epoch} = \frac{\text{size of training set}}{\text{batch size}} = \frac{|T|}{|T_n|}, \tag{8.12}$$

when assuming that we do not change the batch size T_n within an epoch.

The notion of epoch is first useful when comparing the performance of two implementations of the GD or SGD algorithms. To see which algorithm is more effective, for example, one would like to compare the computational cost to achieve some given level of precision. However, when comparing two implementations of SGD with different batch sizes, we cannot just use the number of iterations/steps. This is because for the implementation with smaller batch size, each iteration is faster than in the implementation with larger batch size.

Using the number of epochs instead of steps allows for a more direct comparison. Observe that if an SGD algorithm runs for one epoch, it has selected a total of $|T|$ elements of the training sets when choosing random batches. Therefore, two algorithms with different batch sizes will still perform the same number of calculations of $\nabla \ell(s, \boldsymbol{a})$ over an epoch. Similarly, an SGD algorithm which has run for one epoch will have taken the same number of gradients as a GD algorithm which has run for one iteration over the entire training set, and may therefore be compared fairly.

In some settings, epochs also play a more concrete role in the execution of the algorithm depending on how each batch is randomly selected. A thorough discussion of how to construct batches is beyond the scope of these notes, but to remain at an elementary level, one may distinguish between two cases:

- One selects randomly the objects in the batch T_n independently of the choices made in previous batches. This implies that the same object may be selected again in consecutive batches.
- One may instead use entirely distinct batches. This can be done by imposing the condition that we cannot take the same object twice in different batches within the same epoch.

In the second case, we can then guarantee that each object will have been used exactly once within each epoch of the algorithm.

8.5 Weights

A further question when implementing GD or SGD algorithms is whether to weight objects equally or to give more importance to some of them.

There are naturally two occurrences in an SGD algorithm where distinct objects' weights could play a role. First of all, when randomly selecting a subset of T as our batch for each iteration, we could decide that some objects should be more likely to appear in a batch than others.

Similarly, we could decide to weight differently the various contributions in the loss function: Instead of (8.8), one could use

$$L(\boldsymbol{a}) = \sum_{s \in T} \nu(s)\, \ell(s, \boldsymbol{a}),$$

where $\nu(s)$ is the weight of the object s and those weights are normalized so that $\sum_{s \in T} \nu(s) = 1$.

Weighting directly the loss function does not involve the notion of batch and can also be used in deterministic GD algorithms. For SGD weighting the loss function or weighting the likelihood of selecting an object may have similar effects: In both cases, the gradient $\nabla\ell(s, \boldsymbol{a})$ will be used more for some objects s than others.

As is often the case, there is no general rule on whether to weight equally all objects or not. For example, if a training set contains objects from two classes, but there are significantly more objects from one class than from the other, then one may weigh the objects from the underrepresented class so on average each batch contains as many objects from one class as from the other.

8.6 Choosing the batch size through a numerical example

To conclude our discussion of GD algorithms, we illustrate through a numerical example how one may choose in practice an appropriate batch size for SGD. We consider a relatively simple setting where we can calculate the analytical solution and use it when evaluating the numerical approximations.

We focus on a regression problem to find a function $f : \mathbb{R} \to \mathbb{R}^n$ that best fits our data. The dataset consists of N pairs $(x_i, \boldsymbol{y}_i)$, where each $x_i \in \mathbb{R}$ and $\boldsymbol{y}_i \in \mathbb{R}^n$ for $i = 1, \ldots, N$. We want to approximate $f(x)$ by $\tilde{f}(x, \boldsymbol{a})$, where $\boldsymbol{a} \in \mathbb{R}^\mu$ is a vector of parameters. Specifically, we choose $\tilde{f}$ to have the form

$$\tilde{f}(x, \boldsymbol{a}) = (\tilde{f}_1(x, \boldsymbol{a}), \ldots, \tilde{f}_n(x, \boldsymbol{a})), \tag{8.13}$$

where each $\tilde{f}_i(x, \boldsymbol{a})$ is a polynomial of degree d in x. Since each of the n polynomials of degree d has $d + 1$ coefficients (parameters), the parameter vector $\boldsymbol{a}$ has $n(d + 1)$ components, that is, the dimension μ of the parameter space is $n(d + 1)$.

We can first find analytically the parameters $\boldsymbol{a}$ that yield the best fit $\tilde{f}$ by minimizing the mean square loss (6.2) from Section 6.1,

$$L(\boldsymbol{a}) = \frac{1}{N} \sum_{i=1}^{N} \|\tilde{f}(x_i, \boldsymbol{a}) - \boldsymbol{y}_i\|^2. \tag{8.14}$$

Observe that each polynomial $\tilde{f}_i(x, \boldsymbol{a})$ depends linearly on its coefficients. Therefore *$\tilde{f}(x_i, \boldsymbol{a})$ depends linearly on $\boldsymbol{a}$.* It is convenient to rearrange the vector $\boldsymbol{a}$ as an $n \times (d + 1)$ matrix A where the ith row of A contains the coefficients of the polynomial $\tilde{f}_i$. Then $L(\boldsymbol{a}) = L(A)$, where

$$L(A) = \frac{1}{N} \sum_{i=1}^{N} \|A\hat{\boldsymbol{x}}_i - \boldsymbol{y}_i\|^2, \tag{8.15}$$

with $\hat{\boldsymbol{x}}_i = (1, x_i, x_i^2, \ldots, x_i^d) \in \mathbb{R}^{d+1}$ so that $A\,\hat{\boldsymbol{x}}_i$ is a vector which contains the values of the n polynomials $\tilde{f}_i$ at the point x_i.

The gradient $\nabla L(A)$ is now explicitly given by

$$\nabla L(A) = \frac{2}{N}\sum_{i=1}^{N}(A\hat{\boldsymbol{x}}_i - \boldsymbol{y}_i)\hat{\boldsymbol{x}}_i^T. \tag{8.16}$$

Recall that the minimizer $A_{\min}$ of $L(A)$ satisfies $\nabla L(A_{\min}) = 0$. Therefore, $A_{\min}$ may be calculated from (8.16) as

$$A_{\min} = \left(\sum_{i=1}^{N}\boldsymbol{y}_i\hat{\boldsymbol{x}}_i^T\right)\left(\sum_{i=1}^{N}\hat{\boldsymbol{x}}_i\hat{\boldsymbol{x}}_i^T\right)^{-1}, \tag{8.17}$$

where $\boldsymbol{y}_i\hat{\boldsymbol{x}}_i^T$ and $\hat{\boldsymbol{x}}_i\hat{\boldsymbol{x}}_i^T$ are matrices with $(\boldsymbol{y}_i\hat{\boldsymbol{x}}_i^T)_{jk} = (\boldsymbol{y}_i)_j(\hat{\boldsymbol{x}}_i)_k$ and $(\hat{\boldsymbol{x}}_i\hat{\boldsymbol{x}}_i^T)_{jk} = (\hat{\boldsymbol{x}}_i)_j(\hat{\boldsymbol{x}}_i)_k$, respectively. The minimizing matrix $A_{\min}$ of $L(A)$ corresponds to a minimizing vector $\boldsymbol{a}_{\min}$ of $L(\boldsymbol{a})$.

One may expect that since it is possible to explicitly calculate the minimizer $\boldsymbol{a}_{\min}$, there is no need to use numerical methods like SGD to approximate it. However, first of all, if N, n, and d are large, then calculating (8.17) is computationally expensive, particularly due to the need to invert the matrix $\sum_{i=1}^{N}\hat{\boldsymbol{x}}_i\hat{\boldsymbol{x}}_i^T$. In this case it may still be faster to use SGD particularly because the batch size may be much smaller than N.

Second, we also stress again that in more complex examples calculating an exact analytical solution may not be feasible. This example was specifically chosen so that the analytical solution exists and can be compared to numerical implementations of SGD.

We now turn to the implementation of SGD algorithms to minimize (8.14) for $N = 2^{13}$, $n = 64$, and $d = 3$. Our objective is to show how the batch size in the SGD algorithm changes the rate of convergence. We run SGD 12 times with batch sizes 2^k for $k = 0, 1, 2, \dots, 11$. At each run, we train for 200 epochs, recording the loss and computation time (in seconds) at the end of each epoch.

Let $\hat{\boldsymbol{a}}$ be the minimizer calculated by (8.17) and let $L_{\min} = L(\hat{\boldsymbol{a}})$ be the global minimum of the loss function. We introduce the *relative loss*

$$\bar{L}(\boldsymbol{a}) = \frac{L(\boldsymbol{a})}{L_{\min}} - 1, \tag{8.18}$$

which shows how close $L(\boldsymbol{a})$ is to $L_{\min}$.

Figs. 8.3a–8.3c show how the relative loss changes with the epoch number for each batch size. We observe from Fig. 8.3 that for each batch size b, the loss function first rapidly decreases with the number of epochs before roughly leveling out with rapid oscillations about some value of relative loss $\bar{L}_b$. Observe that $\bar{L}_b$ decreases as b increases, but the number of epochs required to reach $\bar{L}_b$ increases with b. Thus, larger batch sizes lead to higher precision than smaller batch sizes, but may take longer to level out (see in particular Fig. 8.3c).

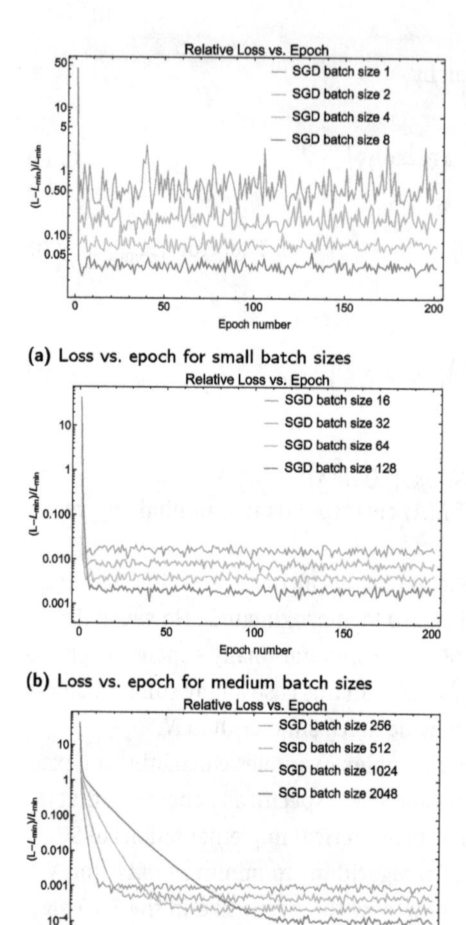

(a) Loss vs. epoch for small batch sizes

(b) Loss vs. epoch for medium batch sizes

(c) Loss vs. epoch for large batch sizes

Figure 8.3: Log plots of loss vs. epoch for several batch sizes of SGD and deterministic GD. Note that the vertical axis is not actual mean square loss, but the relative difference between loss and the absolute minimum loss: $\bar{L} = (L - L_{\min})/L_{\min}$. Observe that each curve decreases rapidly before approximately reaching a minimum and then rapidly oscillating near this minimum.

Fig. 8.4 shows the amount of computation time[1] (in seconds) required for the SGD algorithm to reach a relative loss value of 0.04 for each batch size. Observe that (i) the smallest batch sizes never even reach a relative loss value of this level, (ii) large batch sizes take more time, and (iii) the optimal batch size (shortest time to reach this loss) appears to be about 32.

1 The code for this example was written in Mathematica 11.0 and was run on an 11th Gen Intel(R) Core(TM) i7-1165G7 @ 2.80 GHz.

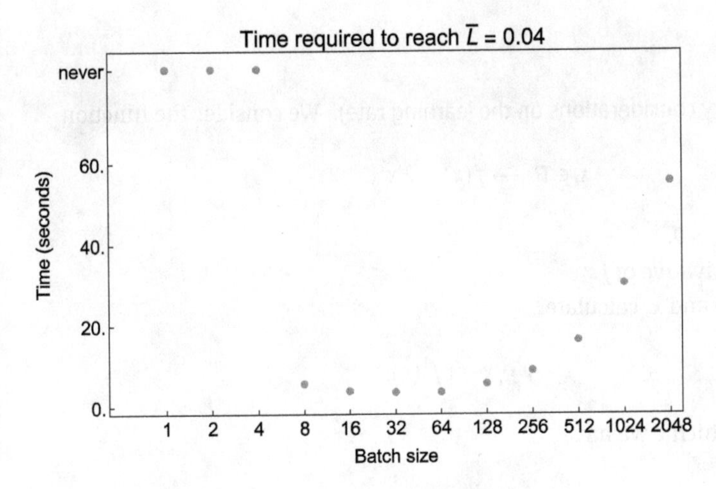

Figure 8.4: The amount of time required to reach the loss.

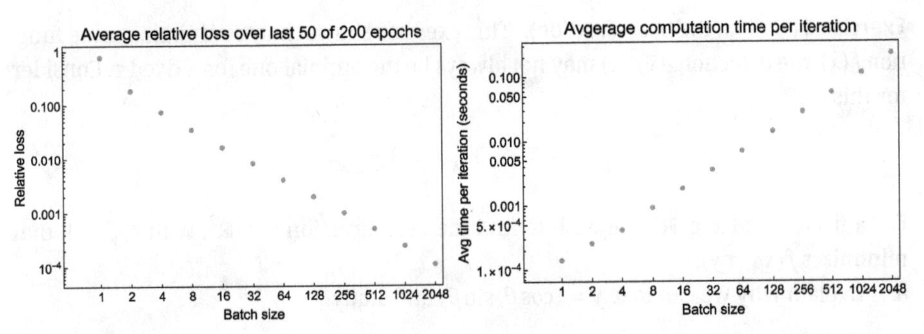

(a) Average relative loss over last 50 epochs (out of 200)

(b) Average computation time per iteration

Figure 8.5: Average time and relative loss.

Fig. 8.5a supports our previous findings. Fig. 8.5a shows the relative loss averaged over the last 50 epochs of training. Clearly, larger batch sizes lead to smaller relative loss. Fig. 8.5b shows the average computation time per iteration, which is, as expected, linear in terms of the batch size.

To summarize:

- SGD algorithms with small batch size b_{small} very quickly reach a higher relative loss $\bar{L}_{b_{\text{small}}}$ and do not really improve afterwards.
- SGD algorithms with large batch size b_{large} reach the relative loss $\bar{L}_{b_{\text{large}}}$ more slowly, that is, they requires more epochs, but the resulting loss $\bar{L}_{b_{\text{large}}}$ is smaller ($\bar{L}_{b_{\text{large}}} < \bar{L}_{b_{\text{small}}}$).
- For a fixed value of relative loss $\bar{L}_0$, there is an optimal batch size b_0, which is not so small that SGD does not reach $\bar{L}_0$ and not so large that SGD takes many epochs to reach $\bar{L}_0$. In our example, $b_0 \approx 32$, as follows from Fig. 8.4.

8.7 Exercises

Exercise 1 (Elementary considerations on the learning rate). We consider the function

$$x \in \mathbb{R} \longrightarrow f(x) = \lambda x^2,$$

for a fixed constant $\lambda > 0$.

a. Calculate the derivative of f.
b. For a given $\tau > 0$ and x, calculate

$$y = x - \tau f'(x).$$

c. Determine for which τ we have

$$f(y) < f(x).$$

Exercise 2 (Finding the best direction). This exercise shows that to minimize some function $f(x)$, the direction $-\nabla f(x)$ may not always be the optimal one for a fixed τ. Consider for this

$$x \in \mathbb{R}^2 \longrightarrow f(x) = x_1^2 + 2x_2^2.$$

For a fixed τ and $x \in \mathbb{R}^2$, we wish to find the best direction $y \in \mathbb{R}^2$, with $\|y\| = 1$, that minimizes $f(x - \tau y)$.

a. Explain why we can take $y = (\cos\theta, \sin\theta)$ and study

$$g(\theta) = f(x_1 - \tau \cos\theta,\ x_2 - \tau \sin\theta).$$

b. Calculate $g'(\theta)$ and show that the minimum is obtained for

$$y_1 = \frac{x_1 y_2}{2x_2 - \tau y_2}.$$

c. Calculate $\nabla f(x)$.
d. Show that we can take y parallel to $\nabla f(x)$ if $\tau = 0$ but that y and $\nabla f(x)$ are not collinear if $\tau > 0$.

Exercise 3 (Oscillations in gradient descent algorithms). We are using a GD algorithm with constant learning rate on the function

$$x \in \mathbb{R} \longrightarrow f(x) = (\sin 10x)^2.$$

Naturally, f has a minimum at $x = 0$.

a. Explain why we cannot expect in general to have convergence to the minimum at $x = 0$ if the learning rate $\tau > 1/19$.

b. Verify this result numerically by testing various learning rates τ by writing your own code.

Exercise 4 (The role of weights in the limit of GD algorithms). We consider the following quadratic loss function, corresponding to a simple case with one parameter and three objects:

$$\alpha \in \mathbb{R} \longrightarrow L(\alpha) = (\alpha - s_1)^2 + (\alpha - s_2)^2 + \nu\,(\alpha - s_3)^2,$$

where $\nu > 0$ is a weight that can be chosen. The values s_1, s_2, s_3 correspond to the three objects, and we assume that

$$s_1 = 0, \quad s_2 = 0.01, \quad s_3 = 1.$$

a. Calculate $L'(\alpha)$.
b. For each ν find the value α_ν that minimizes $L'(\alpha)$.
c. Explain how α_ν changes in terms of ν and interpret this in terms of how well classified each object is.

Exercise 5 (Convergence of gradient descent for non-convex functions). We consider a non-convex smooth function $f : \mathbb{R}^n \to \mathbb{R}_+$. We assume that all second-order partial derivatives of f are bounded and recall that this implies, through Taylor expansion, the following inequality: There exists $\kappa > 0$ such that for any α and β in $\mathbb{R}^n$,

$$\left|f(\beta) - f(\alpha) - (\beta - \alpha) \cdot \nabla f(\alpha)\right| \le \kappa\,\|\alpha - \beta\|^2. \tag{8.19}$$

We implement a GD algorithm with non-constant learning rate that consists in calculating the sequence α_k with

$$\alpha_{k+1} = \alpha_k - \tau_k\,\nabla f(\alpha_k).$$

a. Use (8.19) to show that provided $\tau_k\,\kappa \le 1/2$, we have $f(\alpha_{k+1}) \le f(\alpha_k)$ with in particular

$$f(\alpha_{k+1}) \le f(\alpha_k) - \tau_k\,\|\nabla f(\alpha_k)\|^2.$$

b. Deduce that if $\tau_k\,\kappa \le 1/2$, then

$$\sum_{k=1}^{\infty} \tau_k\,\|\nabla f(\alpha_k)\|^2 < \infty.$$

c. Deduce further that provided $\sum_{k=1}^{\infty} \tau_k = +\infty$, one has $\|\nabla f(\alpha_k)\| \to 0$ as $k \to \infty$.

d. We finally assume that f has only a finite number of critical points: There only exist finitely many points $y_1, \ldots, y_N$ such that $\nabla f(y_i) = 0$. Conclude that the sequence a_k must converge to one of the y_i.

Exercise 6 (Some advantages of stochastic gradient descent). Consider a finite number of differentiable functions $\ell_i(a) : \mathbb{R}^n \to \mathbb{R}$, each corresponding to a given object s_i, $i = 1 \ldots N$, in the training set T. We attempt to minimize the full loss function

$$L(a) = \frac{1}{N} \sum_{i=1}^{N} \ell_i(a)$$

through an SGD algorithm with constant learning rate and batch size S.

a. Recall the definition of SGD.
b. We choose a batch size of one and assume that the random sequence a_k obtained from SGD converges almost surely to some $\bar{a}$. Prove that

$$\nabla \ell_i(\bar{a}) = 0 \quad \text{for all } i.$$

Explain why this is better than if we had convergence for deterministic GD.
c. Show that the same is true for any batch size.

9 Backpropagation

The training of deep neural networks (DNNs) requires to efficiently compute $\nabla L(\alpha)$, where the vector of parameters $\alpha = (\alpha_k) \in \mathbb{R}^\mu$ has large dimension $\mu \gg 1$; currently for the largest DNNs, μ can be more than half a trillion! Calculating individually all partial derivatives for each one of these μ parameters is time consuming, even for high-performance computers. Therefore, techniques for expediting the calculation of derivatives are valuable tools for training DNNs.

One of the most effective of these techniques is *backpropagation*, which takes advantage of the fact that $L(\alpha)$ is not an arbitrary function of many parameters. On the contrary, the dependence of $L(\alpha)$ on the parameters α is determined by the compositional structure of a DNN; see Definition 4.1.3 and Fig. 4.4. Then computing $\nabla L(\alpha)$ using the compositional structure of (4.8) leads to applying the chain rule to each layer iteratively. The key observation here is that in this iterative process from layer to layer of the DNN, most of the derivatives appear more than once. The computational savings come from eliminating repetitive computation of such derivatives.

9.1 Computational complexity

In order to understand how one algorithm is more efficient than another, we need to define how to "measure" efficiency. To this end, we briefly review the basic concepts of computational complexity. Roughly speaking, the notion of complexity corresponds to the total number of operations (e. g., addition and multiplication) required to execute a computational algorithm. The following example in an everyday context gives an idea of computational complexity.

Example 9.1.1 (Comparing lists of names). Suppose you are given two lists of names A and B that you have to compare, for example a list of the names of students registered to Calculus 1 and a list of the names of students registered to Linear Algebra. The goal is to mark all the names that are on both list A and list B. For simplicity, the computational complexity in this example will be the total number of times that we read a name, e. g., reading one name twice counts as two operations, and we assume that lists A and B both have n names.

We are going to consider three cases of how the lists can be ordered.

(i) The names in both lists were entered in alphabetical order.

(ii) The names in both lists were entered in random order, for example by writing the names of the students in the order in which they registered for the course.

(iii) The names in one list, for example list A, are in alphabetical order, while the names in list B are in random order.

https://doi.org/10.1515/9783111025551-009

Intuitively the first case should be the easiest and the second the most time consuming. We can quantify this by evaluating the computational complexity in each of the cases above. In the first case, we can go sequentially at the same time through each list marking the names in list A that are not in list B. Thus we read both lists once for a total of $n + n = 2n$ names read.

In the second case, we can go through the names in list B in the order in which they are entered but for each name we need to go through list A and check whether the corresponding name is in list A. This may require going through list A one-by-one for each name in list B. To be more mathematically precise, we can number the names on list B from 1 to n; for each i between 1 and n, we denote by p_i the number of names we must read in list A before finding name i (or we read all the names in list A and deduce that name i is not in this list, in which case $p_i = n$). The total cost is then

$$\sum_{i=1}^{n}(1 + p_i), \tag{9.1}$$

where $(1 + p_i)$ corresponds to first reading a name in list B (which has cost 1) and then reading p_i names in list A (which has cost p_i).

The exact total depends on the exact values of p_i. However, by considering the best- and worst-case random orderings, we can easily provide lower and upper bounds as follows. In the best-case scenario, our two lists contain the same elements with perhaps different ordering, and hence each p_i has to be between 1 and n and must be different, giving us a lower bound (see (9.2)). In the worst-case scenario, $p_i = n$ for all i because lists A and B share no elements. Hence we obtain

$$\frac{n(n+3)}{2} = \sum_{j=1}^{n}(1 + j) \le \sum_{i=1}^{n}(1 + p_i) \le \sum_{j=1}^{n}(1 + n) = n(n+1). \tag{9.2}$$

The third case is somewhat more complex and the computational complexity here ends up being intermediate between the fastest case 1 and the slowest case 2.

We explain now an algorithm, called the *binary search algorithm*, which uses the alphabetical structure of list A to efficiently mark all names on list A that are also on list B. The algorithm works as follows: For any given name #i in list B we check if this name is in list A or not. To this end, we begin by looking at the name in the middle of list A. If this name matches our desired name #i, then our algorithm terminates. Otherwise, since A is in alphabetical order, we will know if name #i is in the first half of A or in the second half, in the same way that one figures out whether one has passed a word in the dictionary. We then repeat this process on the half of the list that may contain name #i until we find it in list A (or determine that name #i is not in list A).

In this case, for each student in list B we perform a binary search algorithm to see if their name appears in list A. Each binary search algorithm requires at most $\log_2(n)$ operations. To understand why, we consider the maximum number of times that a list of n elements can be split in half. In the first step, we cut the second list in half, resulting in $\frac{n}{2}$ names left in an updated list. In the second step we once again cut the new updated list in half, resulting in $\frac{n}{4}$ names in the updated list. We continue doing this until we have one name left, so we need to find x such that $2^x = n$. Thus $x = \log_2(n)$, so we divide list A into equal halves a maximum of $\log_2(n)$ times. Finally, because there are n names in list B, each of which requires this binary search algorithm, the total maximum number of operations we perform is $n + n\log_2(n)$.

As one can see, the cost in this case, $n + n\log_2(n)$, is much lower than in the second case but still higher than in the first case. Of course it is possible to be more efficient in the third case, for example by marking where the first name starting with letter A, B, C, etc., is located.

When it comes to computational complexity we are typically concerned about how the algorithms behave for a large number of operations. However, it is often not clear how to compare exactly individual operations: The time to perform a multiplication vs. an addition may depend on the processor for example. It may also be tedious or not simple to calculate the exact number of operations, as in the example above. For this reason, we often wish to focus on the order of magnitude without considering precise numerical constants. The following definition provides a rigorous mathematical meaning to this idea.

Definition 9.1.1 ($O(\cdot)$ Landau symbol). Consider two functions $f, g : \mathbb{N} \to \mathbb{R}$. If there exist a constant $C > 0$ and an integer N such that

$$|f(n)| \leq C|g(n)| \tag{9.3}$$

for all $n > N$, then f is said to be of at most order g as $n \to \infty$. We denote this by

$$f(n) = O(g(n)) \quad \text{as } n \to \infty. \tag{9.4}$$

A graphical example of $f(n) = O(g(n))$ where inequality (9.3) is satisfied for large n is shown in Fig. 9.1. Note that $g(n)$ may in fact be much larger than $f(n)$ as $n \to \infty$.

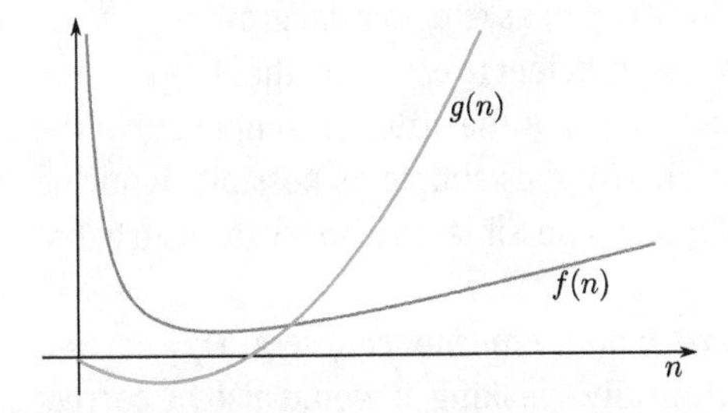

Figure 9.1: Example of $f(n) = O(g(n))$.

The following proposition shows how $O(\cdot)$ can be determined using a limit.

Proposition 9.1.1. *Let $f, g : \mathbb{N} \to \mathbb{R}$. If $\lim_{n\to\infty} f(n)/g(n)$ exists and is finite, then $f(n) = O(g(n))$ as $n \to \infty$. Moreover, if the limit is non-zero, then $g(n) = O(f(n))$ as $n \to \infty$.*

Proof. Suppose $\lim_{n\to\infty} f(n)/g(n) = \ell < \infty$. Then there exists N such that if $n > N$, then $|f(n)/g(n) - \ell| < 1$, which leads to the following sequence of inequalities:

$$\left|\frac{f(n)}{g(n)} - \ell\right| < 1 \tag{9.5}$$

$$\Rightarrow \left|\frac{f(n)}{g(n)}\right| - |\ell| < 1 \text{ (triangle inequality)} \tag{9.6}$$

$$\Rightarrow \frac{|f(n)|}{|g(n)|} < 1 + |\ell| \tag{9.7}$$

$$\Rightarrow |f(n)| < (1 + |\ell|)|g(n)|, \tag{9.8}$$

for all $n > N$. Letting $C = (1+|\ell|)$, we observe that f, g satisfy (9.3) for all $n > N$. Therefore, $f(n) = O(g(n))$ as $n \to \infty$.

If $\ell \neq 0$, then we observe that $\lim_{n\to\infty} g(n)/f(n) = 1/\ell$, so we may repeat the above argument to see that $g(n) = O(f(n))$ as $n \to \infty$. □

It is sometimes easier to calculate a limit than to show that inequality (9.3) holds directly, which is when Proposition 9.1.1 is useful.

Example 9.1.2. Let $f(n) = n$ and $g(n) = n^2$. The function $f(n)$ is of order $g(n)$ because $n \leq n^2$ (set $C = 1$ for $n \geq 1$).

However, $g(n)$ is not of order $f(n)$ because

$$\lim_{n\to\infty} \frac{n^2}{n} = \infty. \tag{9.9}$$

There are some common orders that are worth mentioning. For instance,

$$O(1) \leq O(\log n) \leq O(n^s) \leq O(\exp n), \quad s \in \mathbb{R}^+, n \to \infty. \tag{9.10}$$

We often refer to $O(1)$ functions as bounded in n as $n \to \infty$, to $O(\log n)$ as logarithmic growth, to $O(n^s)$ as polynomial growth, and to $O(\exp n)$ as exponential growth.

If $f(n)$ gives the number of operations that is sufficient to complete the computation and $g(n)$ is such that $f(n) = O(g(n))$, we can also say that the order of complexity of the computation is $O(g(n))$. Generally, we try to choose g as simple as possible, ignoring constant factors and smaller-order terms. Whenever possible we also try to ensure that $g(n) = O(f(n))$.

For example, if $f(n) = \pi n^2 + 3n$, then we would typically choose $g(n) = n^2$ and explain that the algorithm is of order n^2. Technically speaking, it would also be correct to write $f(n) = O(n^3)$, but this is obviously much less useful.

Clearly, algorithms whose computational complexity grows linearly, namely $O(n)$, take much less time to execute than algorithms whose complexity grows quadratically, that is, $O(n^2)$.

Hereafter, counting the number of operations will play a key role in determining the order of computational complexity. As mentioned above, depending on the computing platform, some elementary operations may take proportionally longer than others. To keep the discussion simple and independent of the platform, we however count all those operations identically. This leads to the following definition.

Definition 9.1.2. In a algorithm, an *operation* is any one of the following:
- addition,
- subtraction,
- multiplication,
- division,
- exponentiation,
- comparison of the values of two numbers (e. g., checking whether $x < y$, $x, y \in \mathbb{R}$).

Now we apply Definitions 9.1.1 and 9.1.2 to two simple computations.

Example 9.1.3 (Computational complexity of the dot product). Let $a = (a_1, \ldots, a_n)$, $b = (b_1, \ldots, b_n) \in \mathbb{R}^n$ be two vectors. The dot product of a and b is given by

$$a \cdot b = a_1 b_1 + a_2 b_2 + \cdots + a_n b_n. \tag{9.11}$$

First, we need to perform n multiplications: $a_1 b_1, a_2 b_2, \ldots, a_n b_n$. Then, we need to perform $n-1$ additions: $a_1 b_1 + a_2 b_2$, then $(a_1 b_1 + a_2 b_2) + a_3 b_3$ and so on. The total number of operations according to our definitions is $n + (n-1) = 2n - 1$. Since

$$\lim_{n \to \infty} \frac{2n-1}{n} = 2, \tag{9.12}$$

the computational complexity of the dot product in $\mathbb{R}^n$ is $O(n)$.

Recall the algorithm to multiply two matrices. Since it involves additions and multiplications, we can find its complexity in a similar way.

Example 9.1.4 (Computational complexity of matrix multiplication). Let $A = (a_{ij})$, $B = (b_{ij}) \in \mathbb{R}^{n \times n}$ be two $n \times n$ matrices. Their product is given, componentwise, by

$$(AB)_{ij} = \sum_{k=1}^{n} a_{ik} b_{kj}, \quad i, j \in \{1, \ldots, n\}. \tag{9.13}$$

First, fix a component of the resulting matrix. In order to compute this component we perform $2n - 1$ operations (n multiplications and $n - 1$ additions), since it is the dot product of a row of A and a column of B. This occurs with each of the n^2 components. Therefore, we require $n^2(2n-1)$ operations. Since

$$\lim_{n\to\infty} \frac{n^2(2n-1)}{n^3} = 2, \tag{9.14}$$

the complexity of multiplication of $n \times n$ matrices is $O(n^3)$.

It is actually possible to do better in terms of computational complexity, since the pioneering work [50]. The computational complexity of matrix multiplication has been recently reduced to approximately $O(n^{2.3728596})$ in [1]. See also the recent work [12]. Strassen algorithms [50] are mostly only used in practice for very large matrices. Moreover, on modern computers, memory access or parallelization often proves to be more critical for the final computing time.

It will be useful to note that multiplying a $k \times m$ rectangular matrix by an $m \times n$ rectangular matrix requires

$$O(kmn) \text{ operations.} \tag{9.15}$$

Observe that if $k = m = n$, then the matrices are square, and this agrees with the $O(n^3)$ complexity calculated above.

Finally, we note that while in the above examples the operations were multiplications and additions, in backpropagation other operations are involved, such as determining whether a number is positive or negative (e. g., for ReLU).

9.2 Chain rule review

There is a standard definition of the chain rule that can be found in calculus textbooks (see Appendix). Here we present two simple examples for the single and multivariable cases.

Example 9.2.1. Let $f : \mathbb{R} \to \mathbb{R}$, $g : \mathbb{R} \to \mathbb{R}$ be defined by $f(x) = \cos(x)$, $g(x) = e^x + 2x$. Find

$$\frac{d}{dx} f \circ g(x)\Big|_{x=0}. \tag{9.16}$$

The chain rule for one variable functions gives

$$\frac{d}{dx} f \circ g(x)\Big|_{x=0} = \frac{d}{dy} f(y)\Big|_{y=g(0)} \cdot \frac{d}{dx} g(x)\Big|_{x=0}. \tag{9.17}$$

Since $f'(y) = -\sin(y)$, $g'(x) = e^x + 2$, and $g(0) = 1$, the desired derivative is

$$\frac{d}{dx} f \circ g(x)\bigg|_{x=0} = -\sin(y)\bigg|_{y=1} \cdot (e^x + 2)\bigg|_{x=0} = -\sin(1) \cdot (1+2) = -3\sin(1). \tag{9.18}$$

Example 9.2.2. Let $f : \mathbb{R}^2 \to \mathbb{R}^2, g : \mathbb{R}^2 \to \mathbb{R}^2$ be functions defined by

$$f(u,v) = (u^2 + 3uv, \exp(u+v)), \tag{9.19}$$

$$g(x,y) = (xy, x^2 + y^2). \tag{9.20}$$

Find all partial derivatives at $(x,y) = (1,0)$

$$\frac{\partial(f \circ g)}{\partial(x,y)}\bigg|_{(x,y)=(1,0)}. \tag{9.21}$$

As a result of the chain rule, this is equal to the following matrix multiplication:

$$\begin{pmatrix} \frac{\partial f_1}{\partial u}(u,v) & \frac{\partial f_1}{\partial v}(u,v) \\ \frac{\partial f_2}{\partial u}(u,v) & \frac{\partial f_2}{\partial v}(u,v) \end{pmatrix}\Bigg|_{(u,v)=g(1,0)} \cdot \begin{pmatrix} \frac{\partial g_1}{\partial x}(x,y) & \frac{\partial g_1}{\partial y}(x,y) \\ \frac{\partial g_2}{\partial x}(x,y) & \frac{\partial g_2}{\partial y}(x,y) \end{pmatrix}\Bigg|_{(x,y)=(1,0)}, \tag{9.22}$$

where $f_1(u,v) = u^2 + 3uv$, $f_2(u,v) = \exp(u+v)$, $g_1(x,y) = xy$, $g_2(x,y) = x^2 + y^2$, and $g(1,0) = (0,1)$. The matrices in (9.22) are called the *Jacobians* or *Jacobian matrices* for the functions $f(u,v)$ and $g(x,y)$.

The first matrix in (9.22) is

$$\begin{pmatrix} 2u + 3v & 3u \\ \exp(u+v) & \exp(u+v) \end{pmatrix}\Bigg|_{(u,v)=(0,1)} = \begin{pmatrix} 3 & 0 \\ e & e \end{pmatrix}, \tag{9.23}$$

and the second matrix in (9.22) is

$$\begin{pmatrix} y & x \\ 2x & 2y \end{pmatrix}\Bigg|_{(x,y)=(1,0)} = \begin{pmatrix} 0 & 1 \\ 2 & 0 \end{pmatrix}. \tag{9.24}$$

Therefore, the desired derivative is given by the product of matrices

$$\begin{pmatrix} 3 & 0 \\ e & e \end{pmatrix} \cdot \begin{pmatrix} 0 & 1 \\ 2 & 0 \end{pmatrix} = \begin{pmatrix} 0 & 3 \\ 2e & e \end{pmatrix}. \tag{9.25}$$

Thus, application of the chain rule leads to matrix multiplication. Since the chain rule is a building block of the backpropagation algorithm, the computational complexity of such multiplication for large n plays an important role in the evaluation of the complexity of backpropagation.

9.3 Diagrammatic representation of the chain rule in simple examples

In this section we introduce a diagrammatic interpretation of the chain rule which is helpful for understanding backpropagation methods for DNN training. This interpretation is also natural when representing DNNs via diagrams; see, e. g., Fig. 4.4.

In the following example, we introduce a toy *backpropagation algorithm* in the simple context of calculating the gradient of a sigmoid-type function.

In this example, the single variable chain rule is not sufficient and the multivariable chain rule is used because several variables are present. However, it is used in a simple form which does not require matrix multiplication of Jacobians. In the next example, the multivariable chain rule involving matrix multiplications is necessary.

Despite the simplicity of Example 9.3.1, it *introduces some of the key ideas of the backpropagation algorithm.*

Example 9.3.1 (Backpropagation algorithm for sigmoid-type functions). We want to compute the gradient of the function $\sigma : \mathbb{R}^4 \to \mathbb{R}$, given by

$$\sigma(x,y,z,w) = \frac{1}{w + \exp(z + \cos(yx^2))}, \tag{9.26}$$

at $(x,y,z,w) = (1/2, \pi, 1, 1)$. As we will see, there is an obvious way to avoid repeating calculations which pre-figures the idea behind backpropagation. This example is a building block for the explanation of backpropagation in the case of the simplest DNN with one neuron per layer; see Proposition 9.4.1 below.

We begin by representing $\sigma(x,y,z,w)$ as a composition of simpler functions. To this end, we define the following functions which cannot be written as a composition of simpler functions. We call such functions *elementary functions.* We have

$$p(x) = x^2, \tag{9.27}$$

$$q(y,p) = y \cdot p, \tag{9.28}$$

$$r(q) = \cos(q), \tag{9.29}$$

$$s(z,r) = z + r, \tag{9.30}$$

$$t(s) = e^s, \tag{9.31}$$

$$u(w,t) = w + t, \tag{9.32}$$

$$v(u) = 1/u. \tag{9.33}$$

Using p, q, r, s, t, u, and v, we may write σ as a composition,

$$\sigma(x,y,z,w) = v(u(w,t(s(z,r(q(y,p(x))))))). \tag{9.34}$$

The purpose of writing σ as a composition of elementary functions is that it is very easy to compute the derivative(s) of each elementary function, and the chain rule provides a way to write the derivative of a composition of elementary functions as a product of the derivatives of the elementary functions. This composition is represented in the diagram in Fig. 9.2.

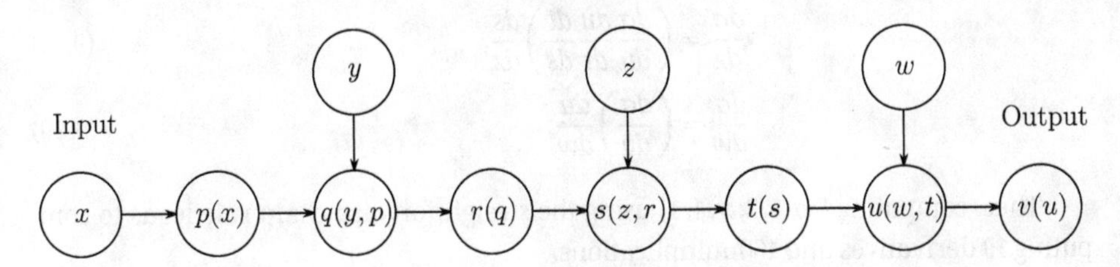

Figure 9.2: Diagrammatic representation of the composition of functions in (9.26).

We now compute $\nabla\sigma(1/2,\pi,1,1)$ using the diagram in Fig. 9.2 in several steps. We will compute the gradient using two methods. The first method is straightforward, but the second reduces the order of complexity. We will show where this reduction in order of complexity comes from.

Step 1. Write the function in each node of each layer of the diagram in terms of the functions in the previous layer. These functions are also called the label of the corresponding node. For example, the first node in the second layer has label $q(y,p) = y \cdot p(x)$, where y and $p(x)$ are labels in the first layer. In this diagram, the elementary functions are written in green in each layer, and they correspond to multiplication, addition, exponentiation, etc.

Step 2. *Forward propagation.* Given values of x, y, z, w, calculate the corresponding values for the functions p, q, r, s, t, u, σ. For the point $(x,y,z,w) = (1/2,\pi,1,1)$, these values are

$$p = 1/4, \quad q = \pi/4, \quad r = \sqrt{2}/2, \quad s = 1 + \sqrt{2}/2, \quad t = e^{1+\sqrt{2}/2}, \tag{9.35}$$

$$u = e^{1+\sqrt{2}/2} + 1, \quad \sigma = \frac{1}{e^{1+\sqrt{2}/2}+1}. \tag{9.36}$$

In the following steps 3–8, we introduce the idea of *backpropagation algorithm* in the context of this example.

Step 3. Compute the derivative function of the functions in each layer with respect to the variable in the previous layer. For example, $\partial t/\partial s = e^s$.

Step 4. Evaluate the derivatives in (9.37)–(9.40) at the point $(1/2,\pi,1,1)$. We now show two ways of evaluating these derivatives.

Step 5. *Naive calculations of derivatives and their products.* We may first naively apply the chain rule from calculus, separately each time for each partial derivative. To compute $\nabla\sigma$ we need to compute all 19 derivatives on the right hand side of (9.37)–(9.40). Furthermore, to apply the chain rule straightforwardly we need an additional 16 multiplications of the derivatives, e. g., in (9.37) we need six operations to compute the product of seven derivatives, leading to 16 total multiplications in (9.37)–(9.40). We have

$$\frac{d\sigma}{dx} = \left(\frac{d\sigma}{du}\frac{du}{dt}\frac{dt}{ds}\frac{ds}{dr}\frac{dr}{dq}\right)\frac{dq}{dp}\frac{dp}{dx}, \tag{9.37}$$

$$\frac{d\sigma}{dy} = \left(\frac{d\sigma}{du}\frac{du}{dt}\frac{dt}{ds}\frac{ds}{dr}\frac{dr}{dq}\right)\frac{dq}{dy}, \tag{9.38}$$

$$\frac{d\sigma}{dz} = \left(\frac{d\sigma}{du}\frac{du}{dt}\frac{dt}{ds}\right)\frac{ds}{dz}, \tag{9.39}$$

$$\frac{d\sigma}{dw} = \left(\frac{d\sigma}{du}\right)\frac{du}{dw}. \tag{9.40}$$

Thus, computing $\nabla\sigma(1/2, \pi, 1, 1)$ using the straightforward chain rule leads to computing 19 derivatives and 16 multiplications.

For the next two steps, recall that complexity comes from two sources: calculation of derivatives and multiplication of derivatives.

Step 6. *Avoiding repetitive calculation of derivatives.* Observe that (9.37)–(9.40) have many repeated terms shown in parentheses. Indeed, in (9.37)–(9.40) there are only ten distinct derivatives. Therefore, the number of derivatives we need to compute can be reduced from 19 in step 5 to ten, which is the total number of edges in Fig. 9.2. The diagram in Fig. 9.2 shows that the savings come from the fact that we traverse each edge only once from any path from right to left.

Step 7. *Clever backward calculation of the product of derivatives.* To complete the application of the chain rule, we need to find the products of the distinct derivatives obtained in step 6. Observe that again many products are repeated in (9.37)–(9.40) and we wish to perform as few multiplications of derivatives as possible. To this end, we view σ as a function of all variables $u, t, s, r, p, q, w, z, y, x$ in the diagram in Fig. 9.2. We then calculate the derivatives of σ with respect to each of these variables starting with u and working *backward*, right to left in the diagram in Fig. 9.2. We reuse any product of derivatives already calculated to avoid repetitions (cf. step 6, where we reuse derivatives already calculated). For example, the product in parentheses in (9.43) was already calculated in (9.42):

$$\frac{\partial\sigma}{\partial u} = -\frac{1}{u^2}, \tag{9.41}$$

$$\frac{\partial\sigma}{\partial t} = \frac{\partial\sigma}{\partial u}\frac{\partial u}{\partial t}, \tag{9.42}$$

$$\frac{\partial\sigma}{\partial s} = \left(\frac{\partial\sigma}{\partial u}\frac{\partial u}{\partial t}\right)\frac{\partial t}{\partial s}. \tag{9.43}$$

This reduces the necessary number of multiplications of derivatives from 20 in step 5 to nine, which is the total number of edges minus one in the diagram in Fig. 9.2. The number of edges is equal to the number of derivatives and the product of k derivatives requires $k - 1$ multiplications.

Step 8. We now use the method described in step 7 to calculate $\nabla\sigma(1/2, \pi, 1, 1)$:

$$\frac{\partial\sigma}{\partial u} = -\frac{1}{u^2} = -\frac{1}{(1+e^{1+\sqrt{2}/2})^2}, \tag{9.44}$$

$$\frac{\partial\sigma}{\partial w} = \frac{\partial\sigma}{\partial u}\frac{\partial u}{\partial w} = -\frac{1}{(1+e^{1+\sqrt{2}/2})^2}\cdot 1, \tag{9.45}$$

$$\frac{\partial\sigma}{\partial t} = \frac{\partial\sigma}{\partial u}\frac{\partial u}{\partial t} = -\frac{1}{(1+e^{1+\sqrt{2}/2})^2}\cdot 1, \tag{9.46}$$

$$\frac{\partial\sigma}{\partial s} = \frac{\partial\sigma}{\partial t}\frac{\partial t}{\partial s} = -\frac{1}{(1+e^{1+\sqrt{2}/2})^2}\cdot e^s = -\frac{e^{1+\sqrt{2}/2}}{(1+e^{1+\sqrt{2}/2})^2}, \tag{9.47}$$

$$\frac{\partial\sigma}{\partial z} = \frac{\partial\sigma}{\partial s}\frac{\partial s}{\partial z} = -\frac{e^{1+\sqrt{2}/2}}{(1+e^{1+\sqrt{2}/2})^2}\cdot 1, \tag{9.48}$$

$$\frac{\partial\sigma}{\partial r} = \frac{\partial\sigma}{\partial s}\frac{\partial s}{\partial r} = -\frac{e^{1+\sqrt{2}/2}}{(1+e^{1+\sqrt{2}/2})^2}\cdot 1, \tag{9.49}$$

$$\frac{\partial\sigma}{\partial q} = \frac{\partial\sigma}{\partial r}\frac{\partial r}{\partial q} = -\frac{e^{1+\sqrt{2}/2}}{(1+e^{1+\sqrt{2}/2})^2}\cdot(-\sin(q)) = \frac{\sqrt{2}e^{1+\sqrt{2}/2}}{2(1+e^{1+\sqrt{2}/2})^2}, \tag{9.50}$$

$$\frac{\partial\sigma}{\partial y} = \frac{\partial\sigma}{\partial q}\frac{\partial q}{\partial y} = \frac{\sqrt{2}e^{1+\sqrt{2}/2}}{(1+e^{1+\sqrt{2}/2})^2}\cdot p = \frac{\sqrt{2}e^{1+\sqrt{2}/2}}{8(1+e^{1+\sqrt{2}/2})^2}, \tag{9.51}$$

$$\frac{\partial\sigma}{\partial p} = \frac{\partial\sigma}{\partial q}\frac{\partial q}{\partial p} = \frac{\sqrt{2}e^{1+\sqrt{2}/2}}{8(1+e^{1+\sqrt{2}/2})^2}\cdot y, \tag{9.52}$$

$$\frac{\partial\sigma}{\partial x} = \frac{\partial\sigma}{\partial p}\frac{\partial p}{\partial x} = \frac{\pi\sqrt{2}e^{1+\sqrt{2}/2}}{8(1+e^{1+\sqrt{2}/2})^2}\cdot 2x = \frac{\pi\sqrt{2}e^{1+\sqrt{2}/2}}{4(1+e^{1+\sqrt{2}/2})^2}. \tag{9.53}$$

In (9.44)–(9.53), we see explicitly the ten distinct derivatives which must be calculated and the nine distinct multiplications which must be made.

In summary, the backward direction of propagation is given by the fact that we need to compute derivatives of σ (rightmost) with respect to variables x, y, z, w (leftmost). There are two sources of savings in this elementary example of backpropagation.

- *First, reusing derivatives already calculated (see step 6).*
- *Second, reusing any product of derivatives already calculated (see step 7).*

Following steps 1–4 and 6–7, we obtain the following gradient for $\sigma(1/2, \pi, 1, 1)$:

$$\sigma_x(1/2, \pi, 1, 1) = \frac{\pi\sqrt{2}e^{1+\sqrt{2}/2}}{4(1+e^{1+\sqrt{2}/2})^2}, \tag{9.54}$$

$$\sigma_y(1/2, \pi, 1, 1) = \frac{\sqrt{2}e^{1+\sqrt{2}/2}}{8(1+e^{1+\sqrt{2}/2})^2}, \tag{9.55}$$

$$\sigma_z(1/2, \pi, 1, 1) = -\frac{e^{1+\sqrt{2}/2}}{(1+e^{1+\sqrt{2}/2})^2}, \tag{9.56}$$

$$\sigma_w(1/2, \pi, 1, 1) = \frac{1}{(1 + e^{1+\sqrt{2}/2})^2}. \quad (9.57)$$

Example 9.3.2. Note that the diagram in Example 9.3.1 is rather simple because every node provides output to only one other node (left to right). However, the exact same steps apply to functions whose diagram contains nodes with multiple outputs. For example, consider the function $f : \mathbb{R}^3 \to \mathbb{R}$, defined by $f(x, y, z) = \sin(xy + xz + yz)$, which has the computation diagram in Fig. 9.3.

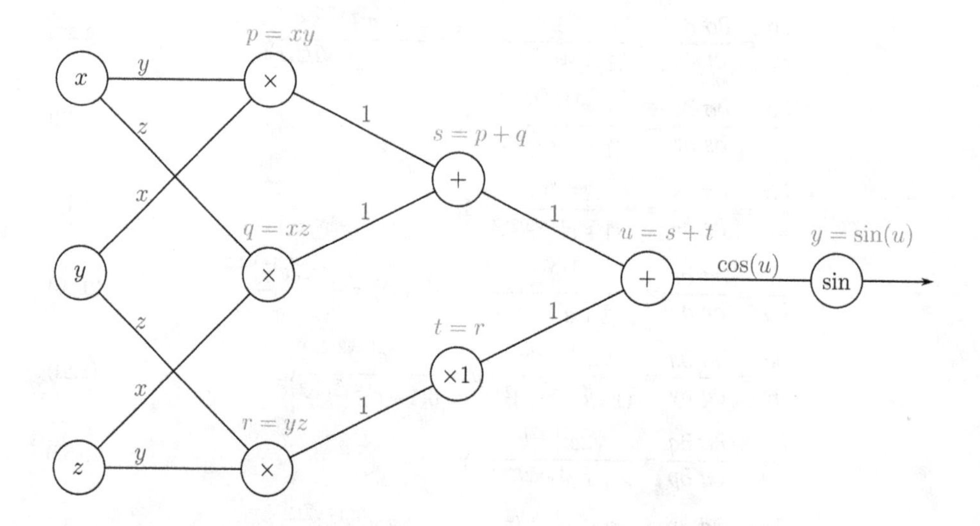

Figure 9.3: Diagram for Example 9.3.2.

9.4 The case of a simple DNN with one neuron per layer

In the previous section we illustrated how backpropagation reduces the number of operations when applying the chain rule to functions of many variables. We used a diagrammatic representation of the chain rule because of the analogy with DNN diagrams. In this section, we extend these backpropagation savings to DNNs and quantify their computational complexity. For simplicity of presentation, we first consider a trivial network with one neuron per layer. The same approach works for general networks (with many neurons per layer), but the calculations are more technical, e. g., by computing gradients with respect to matrices.

If μ is the dimension of the parameter space, we show that the calculation of the gradient of the loss function $L(\boldsymbol{\alpha})$ takes $O(\mu)$ operations (see Definition 9.1.2). Note that if one naively approximates the gradient using finite differences, then the calculation of $L(\boldsymbol{\alpha})$ takes $O(\mu^2)$ operations. Indeed, the gradient in $\boldsymbol{\alpha}$ has μ components. Furthermore, each component requires at least one computation of the artificial neural network (ANN) function, and evaluating the ANN function takes $O(\mu)$ computations because, e. g., each of μ layers requires $O(1)$ computations; see Fig. 9.4. We now contrast this naive approach with the backpropagation algorithm.

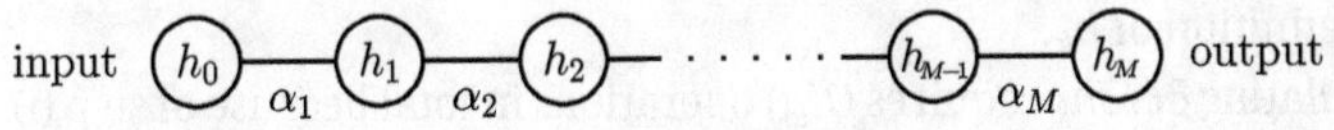

input h_0 —α_1— h_1 —α_2— h_2 — · · · · — h_{M-1} —α_M— h_M output

Figure 9.4: A DNN with many layers and one neuron per layer; h_k is the kth layer function, defined recursively in (9.64).

Proposition 9.4.1. *Let $\phi : \mathbb{R} \times \mathbb{R}^{\mu} \to \mathbb{R}$ be a neural network as in Definition 4.1.3 with M layers, one neuron in each layer, and no biases (i. e., ϕ depends on $\mu = M$ parameters). Let $T \subset \mathbb{R}$ be a training set for ϕ and let $L : \mathbb{R}^{\mu} \to \mathbb{R}$ be a loss function for this DNN of the form (cf. (8.8))*

$$L(\boldsymbol{\alpha}) = \frac{1}{|T|} \sum_{s \in T} \ell(s, \phi(s, \boldsymbol{\alpha})). \tag{9.58}$$

Then for given $\boldsymbol{\alpha} \in \mathbb{R}^{\mu}$ there exists an algorithm called backpropagation which calculates $\nabla L(\boldsymbol{\alpha})$ in $O(|T|\,\mu)$ operations.

Remark 9.4.1. In this proposition, the number of layers M is also the number μ of parameters because there is only one neuron without bias per layer, and therefore only one parameter per layer.

Remark 9.4.2. As we saw before, an example of a specific form of the loss function (9.58) is the mean square approximation of an unknown function $f(x)$ (e. g., exact classifier function) such that

$$\ell(s, \phi(s)) = (\phi(s) - f(s))^2. \tag{9.59}$$

We assume that $f(s)$ is already known for each s, e. g., the result of some measurement, and therefore requires no operations to calculate.

Remark 9.4.3. While Proposition 9.4.1 concerns the gradient of the loss calculated over the entire training set T, the same result holds if one replaces T with a batch $T_n \subset T$ used in, e. g., SGD.

Proof of Proposition 9.4.1. We begin by rewriting formula (9.58) in terms of the layer functions h_k, $k = 1, \ldots, \mu$, to obtain the following formula for $\nabla L(\boldsymbol{\alpha})$:

$$\frac{\partial L}{\partial \alpha_k} = \frac{1}{|T|} \sum_{s \in T} \frac{\partial \ell}{\partial \alpha_k} = \frac{1}{|T|} \sum_{s \in T} \frac{\partial \ell}{\partial h_k} \frac{\partial h_k}{\partial \alpha_k}, \qquad k = 1, \ldots, \mu. \tag{9.60}$$

Next, we note that the number of elements $|T|$ in the training set does not depend on μ, so we may perform all calculations independently for each separate object s. Therefore, it suffices to consider a training set with a single element s. Then

$$\frac{\partial L}{\partial a_k} = \frac{\partial \ell}{\partial h_k}\frac{\partial h_k}{\partial a_k}, \quad k = 1, \ldots, \mu. \tag{9.61}$$

The proof can be given in the following four steps:

(i) Showing that calculating h_k for $k = 1, \ldots, \mu$ requires $O(\mu)$ operations in total because of the recursive definition of h_k.

(ii) Showing that calculating $\partial\ell/\partial h_k$ requires $O(\mu)$ operations in total because of step (i). Otherwise, it would be $O(\mu^2)$ operations. This is the *backpropagation* step when we calculate $\partial\ell/\partial h_k$ when we start with $k = \mu$ and work recursively backward to $k = 1$. The key difference between (i) and (ii) is that h_k is defined by the forward recursive formula (9.64), whereas $\partial\ell/\partial h_k$ is defined via the backward recursive formula (9.68).

(iii) Showing that calculating $\partial h_k/\partial a_k$ for $k = 1, \ldots, \mu$ requires $O(\mu)$ operations in total, again because of step (i) instead of $O(\mu^2)$ operations.

(iv) Showing that calculating $\partial L/\partial a_k$ for $k = 1, \ldots, \mu$ using (9.61) requires μ multiplications of the numbers already calculated in steps (ii) and (iii).

Finally, since each step requires $O(\mu)$ operations, the total number of operations to calculate $\nabla L(\boldsymbol{a})$ for a training set with a single element s is indeed $O(\mu)$ for each object.

Step 1. In this step, we calculate the values of all the layer functions $h_1, \ldots, h_\mu$ in the one-neuron-per-layer network and prove that

the calculation of the values of h_k for $k = 1, \ldots, \mu$ requires $O(\mu)$ operations in total. (9.62)

Hereafter, we refer to h_k as the kth output for brevity.

First we derive a recursive formula for each h_k. The output of the last neuron h_μ is equal to $\phi(s, \boldsymbol{a})$ for object s:

$$\phi(s, \boldsymbol{a}) = \lambda(a_\mu \lambda(a_{\mu-1} \cdots a_2 \lambda(a_1 s) \cdots)), \tag{9.63}$$

where λ is an activation function and a_k is the scalar parameter of the kth layer. Since each layer has one neuron, we write the kth layer function $h_k : \mathbb{R} \to \mathbb{R}$ in the recursive form

$$h_k = \lambda(a_k h_{k-1}), \quad k = 1, \ldots, \mu, \tag{9.64}$$

with

$$h_0 = s \quad \text{and} \quad h_\mu = \phi(s, \boldsymbol{a}). \tag{9.65}$$

Since the output of h_μ is equal to $\phi(s, \boldsymbol{a})$, we may write

$$\phi(s, \boldsymbol{a}) = \phi(h_0, \boldsymbol{a}) = h_\mu \circ h_{\mu-1} \circ \cdots \circ h_1(h_0) \tag{9.66}$$

due to (9.64). Therefore, ϕ and ℓ may be viewed as depending on each h_k, justifying the derivatives in (9.60).

Next, we show that for each k, if h_{k-1} is known, calculating h_k recursively via (9.64) requires $O(1)$ operations. To this end, fix $h_0 \in T$. If h_{k-1} is known, then calculating h_k requires a fixed number of operations that does not depend on μ. For example, if λ = ReLU in (9.64), then calculating h_k requires two operations: one operation to calculate the product $\alpha_k h_{k-1}$ and one operation to determine whether $\alpha_k h_{k-1}$ is positive or negative.

Finally, observe that calculating all the outputs $h_1, \ldots, h_\mu$ requires $O(\mu)$ operations. Indeed, since each h_k requires a fixed number of operations independent of μ, calculating all of the μ outputs $h_1, \ldots, h_\mu$ requires repeating this fixed number of operations μ times, leading to a total of $O(\mu)$ operations. Thus, the claim of step 1 is proved.

Step 2. In this step we prove

$$\text{the calculation of the values of } \frac{\partial \ell}{\partial h_k} \text{ for } k = 1, \ldots, \mu \text{ requires } O(\mu) \text{ operations in total.} \tag{9.67}$$

To this end, we apply the chain rule to $\ell(s, \phi(s, \boldsymbol{\alpha}))$ using (9.63) and (9.64). This yields a formula for $\partial \ell / \partial h_k$ which provides recursion in the *backward direction* as opposed to forward recursion (9.64):

$$\frac{\partial \ell}{\partial h_k} = \frac{\partial \ell}{\partial h_{k+1}} \frac{\partial h_{k+1}}{\partial h_k}, \quad k = 1, \ldots, \mu - 1. \tag{9.68}$$

For $k = \mu$, the derivative $\partial \ell / \partial h_\mu$ is calculated directly since $s = h_0$ and $\phi(s, \boldsymbol{\alpha}) = h_\mu$ as in (9.65), and therefore $\ell(s, \phi(s, \boldsymbol{\alpha})) = \ell(h_0, h_\mu)$. For example, if ℓ is the mean square loss as in (9.59), then there is a function $f : \mathbb{R} \to \mathbb{R}$ such that $\ell(h_0, h_\mu) = (h_\mu - f(h_0))^2$ (see Remark 9.4.2), and

$$\frac{\partial \ell}{\partial h_\mu} = 2(h_\mu - f(h_0)). \tag{9.69}$$

This derivative requires two operations to calculate: one for the subtraction and one for multiplication by 2. Note that h_μ was already calculated in step 1, so no further operations are required to calculate h_μ. Since the total number of operations does not depend on μ, the calculation of $\frac{\partial \ell}{\partial h_\mu}$ takes $O(1)$ operations.

Now we show that the calculation of the remaining derivatives $\partial \ell / \partial h_k$ for $1 \leq k \leq \mu - 1$ also requires $O(1)$ operations for each k. To this end, we apply the *backward* recursive formula (9.68) for $k \leq \mu - 1$. Note that the backward recursion formula (9.68) contains the derivative $\partial h_{k+1} / \partial h_k$, which can be calculated by differentiating (9.64):

$$\frac{\partial h_{k+1}}{\partial h_k} = \lambda'(\alpha_{k+1} h_k) \alpha_{k+1}. \tag{9.70}$$

If h_k has already been calculated (see step 1), then $\frac{\partial h_{k+1}}{\partial h_k}$ can be calculated in $O(1)$ operations: two multiplications and one evaluation of λ'. Similarly, if $\frac{\partial \ell}{\partial h_{k+1}}$ has already been calculated (and since $\partial h_{k+1}/\partial h_k$ requires $O(1)$ operations), the number of operations to calculate $\frac{\partial \ell}{\partial h_k}$ is $O(1)$ due to (9.68). Therefore, since the last derivative $\frac{\partial \ell}{\partial h_\mu}$ has already been calculated in $O(1)$ operations, each of the previous derivatives $\frac{\partial \ell}{\partial h_k}$, $k = 1, \ldots, \mu - 1$, can be calculated in a backward recursive process with $O(1)$ operations in each of the $\mu - 1$ recursive steps, leading to a total of $O(\mu)$ operations to calculate all the derivatives $\frac{\partial \ell}{\partial h_k}$, $k = 1, \ldots, \mu$.

Step 3. Due to (9.60), it remains to show that

$$\textit{the calculation of the values of } \frac{\partial h_k}{\partial \alpha_k} \textit{ for } k = 1, \ldots, \mu \textit{ requires } O(\mu) \textit{ operations in total.} \tag{9.71}$$

Differentiating (9.64), we obtain

$$\frac{\partial h_k}{\partial \alpha_k} = \lambda'(\alpha_k h_{k-1}) h_{k-1}. \tag{9.72}$$

Since h_{k-1} was already calculated in step 1, $\frac{\partial h_k}{\partial \alpha_k}$ can be calculated in $O(1)$ operations for each k: two multiplications and a fixed number of operations to calculate λ'. For example, if λ is ReLU, then

$$\lambda'(x) = \begin{cases} 0 & x \le 0, \\ 1 & x > 0. \end{cases} \tag{9.73}$$

In this case, calculating $\lambda'(\alpha_k h_k)$ requires only one operation: determining whether $\alpha_k h_k$ is positive or negative. This leads to a total of $O(\mu)$ operations to calculate $\frac{\partial h_1}{\partial \alpha_1}, \ldots, \frac{\partial h_\mu}{\partial \alpha_\mu}$.

Step 4. In this step, we use the quantities calculated in steps 2 and 3 to obtain the gradient $\nabla L(\boldsymbol{\alpha})$, and show that it can be done in $O(\mu)$ operations.

From (9.61), we have

$$\frac{\partial L}{\partial \alpha_k} = \frac{\partial \ell}{\partial h_k} \frac{\partial h_k}{\partial \alpha_k}, \quad k = 1, \ldots, \mu. \tag{9.74}$$

Since $\partial \ell/\partial h_k$ and $\partial h_k/\partial \alpha_k$ have already been calculated in steps 2 and 3 for $k = 1, \ldots, \mu$, we only need to perform one multiplication for each $k = 1, \ldots, \mu$, leading to a total of μ operations to compute (9.74).

Since each of steps 1–4 requires $O(\mu)$ operations and since $4 \times O(\mu) = O(\mu)$, we conclude that the calculation of the gradient can be done in $O(\mu)$ operations. □

Important remark

The savings in the backpropagation come from the recursive formulas (9.64) and (9.68). While calculating h_μ requires μ operations, in the process of this calculation, we also calculate h_k for $k = 1, \dots, \mu - 1$ due to (9.64). Similarly, due to (9.68), calculating $\partial\ell/\partial h_1$ requires $O(\mu)$ operations, but $\partial\ell/\partial h_k$ for $k = 2, \dots, \mu$ are calculated along the way. Thus, each of the four steps of the backpropagation algorithm described in the proof above requires $O(\mu)$ operations, but also generates μ pieces of information used in the calculation of the gradient.

9.5 Backpropagation algorithm for general DNNs

In this section, we describe the steps of the backpropagation algorithm applied to DNNs having M layers with no restrictions on their size. We index layers starting from $k = 0$ and denote by n_k the number of neurons in layer #k. We index layer functions starting from 1 so that the kth layer function is $h_k : \mathbb{R}^{n_{k-1}} \to \mathbb{R}^{n_k}$, with n_0 being the number of inputs of the DNN.

From Definition 4.1.2, the layer function is expressed as $h_k(\mathbf{x}) = \bar{\lambda}(A_k\mathbf{x} + \mathbf{b}_k)$, where A_k is an $n_k \times n_{k-1}$ weight matrix, $\mathbf{b}_k \in \mathbb{R}^{n_k}$ is a bias vector (see Fig. 9.5), and $\bar{\lambda}$ is an activation function (recall that $\bar{\lambda} : \mathbb{R}^{n_k} \to \mathbb{R}^{n_k}$ is the componentwise action of a scalar activation function $\lambda : \mathbb{R} \to \mathbb{R}$). Let $\mu_k = n_k(n_{k-1} + 1)$ be the total number of parameters in the kth layer, and let $\alpha_k \in \mathbb{R}^{\mu_k}$ be a vector containing all the parameters in both A_k and $\mathbf{b}_k$ in the kth layer as follows:

$$\boldsymbol{\alpha}_k = ((A_k)_{11}, (A_k)_{12}, \dots, (A_k)_{n_k n_{k-1}}, (\mathbf{b}_k)_1, \dots, (\mathbf{b}_k)_{n_k}). \tag{9.75}$$

Denote by $\mu = \mu_1 + \dots + \mu_M$ the total number of parameters of the DNN and by $\boldsymbol{\alpha} \in \mathbb{R}^\mu$ the vector of all parameters of the DNN:

$$\boldsymbol{\alpha} = (\boldsymbol{\alpha}_1, \dots, \boldsymbol{\alpha}_M). \tag{9.76}$$

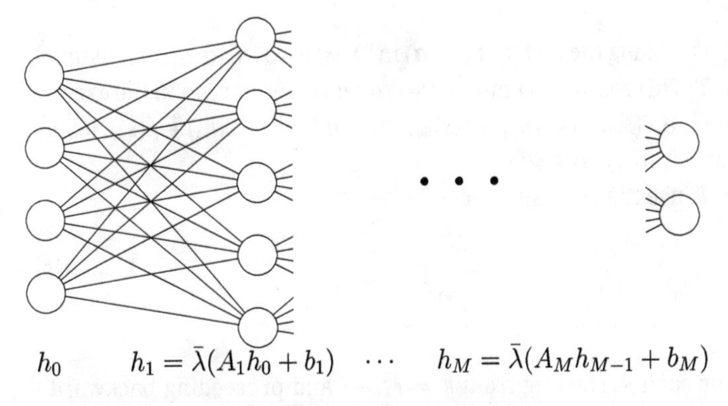

Figure 9.5: Diagrammatic representation for a general DNN.

Let $T \subset \mathbb{R}^{n_0}$ be a training set for the DNN (or a batch of the training set used in SGD), and let $L : \mathbb{R}^{\mu} \to \mathbb{R}$ be a loss function of the form (9.58). We now describe how to implement the backpropagation algorithm to calculate $\nabla L(\boldsymbol{a})$ for a given $\boldsymbol{a} \in \mathbb{R}^{\mu}$:

$$\nabla L(\boldsymbol{a}) = \left(\frac{\partial L}{\partial \boldsymbol{a}_1}, \dots, \frac{\partial L}{\partial \boldsymbol{a}_M} \right), \tag{9.77}$$

where each $\partial L/\partial \boldsymbol{a}_k$ is a vector corresponding to the sum of $\partial \ell/\partial \boldsymbol{a}_k(s,.)$ for each object s in the batch. In turn, each $\partial \ell/\partial \boldsymbol{a}_k(s,.)$ has μ_k components (see Fig. 9.6) given by

$$\underbrace{\frac{\partial L}{\partial \boldsymbol{a}_k}}_{1\times \mu_k \text{ matrix}} = \underbrace{\frac{\partial \ell}{\partial h_k}}_{1\times n_k \text{ matrix}} \underbrace{\frac{\partial h_k}{\partial \boldsymbol{a}_k}}_{n_k\times \mu_k \text{ matrix}} . \tag{9.78}$$

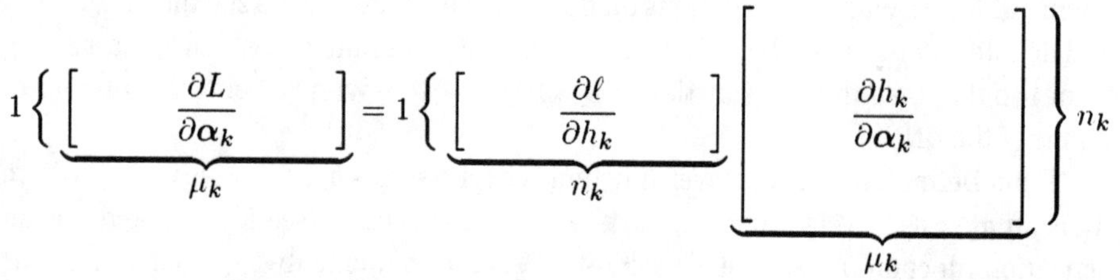

$$1\left\{\underbrace{\left[\quad \frac{\partial L}{\partial \boldsymbol{a}_k} \quad\right]}_{\mu_k}\right. = 1\left\{\underbrace{\left[\quad \frac{\partial \ell}{\partial h_k} \quad\right]}_{n_k}\right. \left.\underbrace{\left[\quad \frac{\partial h_k}{\partial \boldsymbol{a}_k} \quad\right]}_{\mu_k}\right\} n_k$$

Figure 9.6: Dimensions of matrices in (9.80).

0. Forward propagation. Forward propagation (also referred to as a forward pass) is not technically part of the backpropagation algorithm, though it is a necessary preliminary. This is why we denote it by "step 0." To this end, we calculate the value of each layer function h_k, $k = 1, \dots, M$, for each input $s \in T$ using the recursive formula

$$h_k = \bar{\lambda}(A_k h_{k-1} + \boldsymbol{b}_k), \tag{9.79}$$

   where $h_0 = s \in T$.
1. Calculation of $\partial \ell/\partial h_k$. Using the fact that $\phi(s, \boldsymbol{a})$ in (9.58) is equal to h_M, calculate $\frac{\partial \ell}{\partial h_M}$ for each input $s \in T$. This requires at most $O(n_M)$ operations, as seen in the example of mean square loss: If $\ell(s, h_M) = \|h_M - f(s)\|^2$, then $\partial \ell/\partial h_M = 2(h_M - f(s))$, leading to $O(n_M)$ operations as $h_M, f(s) \in \mathbb{R}^{n_M}$.
   Then for each $s \in T$, use the recursive formula

$$\underbrace{\frac{\partial \ell}{\partial h_k}}_{1\times n_k \text{ matrix}} = \underbrace{\frac{\partial \ell}{\partial h_{k+1}}}_{1\times n_{k+1} \text{ matrix}} \underbrace{\frac{\partial h_{k+1}}{\partial h_k}}_{n_{k+1}\times n_k \text{ matrix}} \tag{9.80}$$

   to calculate $\frac{\partial \ell}{\partial h_k}$ for each k, starting from $k = M-1$ and proceeding backward to $k = 1$. The backward recursive formula (9.80) shows how to calculate the derivatives backward starting from the last layer and proceeding to the first, thus allowing us to reuse previously calculated derivatives. The matrices in the right hand side of (9.80) come from the multivariable chain rule (A.8). Specifically,

$$\frac{\partial h_{k+1}}{\partial h_k} = \underbrace{\frac{d\bar{\lambda}}{dx}\bigg|_{x=A_{k+1}h_k+\boldsymbol{b}_{k+1}}}_{n_{k+1}\times n_{k+1}\text{ matrix}} \underbrace{A_k}_{n_{k+1}\times n_k\text{ matrix}}, \tag{9.81}$$

where $\frac{d\bar{\lambda}}{dx}\big|_{x=A_{k+1}h_k+\boldsymbol{b}_{k+1}}$ is a diagonal matrix with the ith diagonal entry being $\lambda'((A_{k+1}h_k + \boldsymbol{b}_{k+1})_i)$. For example, if λ is ReLU, then the ith diagonal entry of $\frac{d\bar{\lambda}}{dx}\big|_{x=A_{k+1}h_k+\boldsymbol{b}_{k+1}}$ is 1 or 0 depending on whether the ith component of $A_{k+1}h_k + \boldsymbol{b}_{k+1}$ is positive or negative.

We first apply straightforwardly the computational complexity formula (9.15) to matrix multiplication in (9.81) without taking into account the specific structure of the matrices (e. g., diagonal). This results in $O(n_{k+1}^2 n_k)$ operations. However, since $\frac{d\bar{\lambda}}{dx}\big|_{x=A_{k+1}h_k+\boldsymbol{b}_{k+1}}$ is a diagonal matrix, the matrix multiplication in (9.81) only requires $O(n_{k+1}n_k)$ operations.

Formula (9.80) involves another $O(n_{k+1}\, n_k)$ operations to estimate the product.

2. Calculation of $\partial h_k/\partial \boldsymbol{a}_k$. Take the derivative of (9.79) in $\boldsymbol{a}_k$ to get

$$\frac{\partial h_k}{\partial \boldsymbol{a}_k} = \underbrace{\frac{d\bar{\lambda}}{dx}\bigg|_{x=A_k h_{k-1}+\boldsymbol{b}_k}}_{n_k\times n_k\text{ matrix}} \underbrace{\frac{\partial(A_k h_{k-1} + \boldsymbol{b}_k)}{\partial \boldsymbol{a}_k}}_{n_k\times n_k(n_{k-1}+1)\text{ matrix}}. \tag{9.82}$$

As in step 1, the matrix $\frac{d\bar{\lambda}}{dx}\big|_{x=A_k h_{k-1}+\boldsymbol{b}_k}$ is diagonal with the ith entry being $\lambda'((A_k h_{k-1} + \boldsymbol{b}_k)_i)$.

For each $k = 1, \dots, M$ and each $s \in T$, calculate the $n_k \times (n_k \cdot n_{k-1})$ matrix[1] $\frac{\partial h_k}{\partial A_k}$ and the $n_k \times n_k$ matrix $\frac{\partial h_k}{\partial \boldsymbol{b}_k}$ using (9.79). Then combine these to form the $n_k \times \mu_k$ matrix $\frac{\partial(A_k h_{k-1}+\boldsymbol{b}_k)}{\partial \boldsymbol{a}_k}$. While not diagonal, this matrix is sparse, meaning that most of its entries are zero. It has the following form:

$$\frac{\partial(A_k h_{k-1} + \boldsymbol{b}_k)}{\partial \boldsymbol{a}_k} = \begin{pmatrix} (h_{k-1})_1 & \cdots & (h_{k-1})_{n_{k-1}} & \cdots & 0 & \cdots & 0 & 1 & \cdots & 0 \\ \vdots & \vdots & \vdots & \ddots & \vdots & \vdots & \vdots & \vdots & \ddots & \vdots \\ 0 & \cdots & 0 & \cdots & (h_{k-1})_1 & \cdots & (h_{k-1})_{n_{k-1}} & 0 & \cdots & 1 \end{pmatrix}. \tag{9.83}$$

Observe from (9.83) that the first n_{k-1} columns of $\frac{\partial(A_k h_{k-1}+\boldsymbol{b}_k)}{\partial \boldsymbol{a}_k}$ have the elements of h_{k-1} in the first row and zeros in all other rows. The next n_{k-1} columns have the ele-

1 Though A_k is a matrix, treat it as a vector in $\mathbb{R}^{n_k n_{k-1}}$ here.

ments of h_{k-1} in the second row. The pattern continues until columns $(n_k-1)n_{k-1}+1$ through $n_k n_{k-1}$ which has the elements of h_{k-1} in the last row. The last n_k columns are the identity matrix. The first $n_k n_{k-1}$ columns of $\frac{\partial(A_k h_{k-1}+\boldsymbol{b}_k)}{\partial \boldsymbol{a}_k}$ represent the derivative of $A_k h_{k-1} + \boldsymbol{b}_k$ with respect to the elements of A_k and the last n_k columns represent the derivative of $A_k h_{k-1} + \boldsymbol{b}_k$ with respect to the elements of $\boldsymbol{b}_k$.

In general the multiplication of an $n_k \times n_k$ matrix by an $n_k \times n_k(n_{k-1}+1)$ matrix as in (9.82) is expected to take $O(n_k^3(n_{k-1}+1))$ operations due to (9.15). However, in (9.82), both matrices are sparse, so the number of operations required is only $O(n_k(n_{k-1}+1))$. Since both factors in (9.82) are sparse, the product $\partial h_k/\partial \boldsymbol{a}_k$ is also sparse, having only one non-zero element in each of its μ_k columns.

3. Calculation of $\partial \ell/\partial \boldsymbol{a}_k$. For each $k = 1, \ldots, M$ and each $s \in T$, calculate $\frac{\partial L}{\partial \boldsymbol{a}_k}$ using

$$\underbrace{\frac{\partial \ell}{\partial \boldsymbol{a}_k}}_{1\times \mu_k \text{ matrix}} = \underbrace{\frac{\partial \ell}{\partial h_k}}_{1\times n_k \text{ matrix}} \underbrace{\frac{\partial h_k}{\partial \boldsymbol{a}_k}}_{n_k\times \mu_k \text{ matrix}}, \tag{9.84}$$

where $\frac{\partial \ell}{\partial h_k}$ is obtained in step 2 and $\frac{\partial h_k}{\partial \boldsymbol{a}_k}$ is obtained in step 3.

4. Sum the derivatives of individual loss $\frac{\partial \ell}{\partial \boldsymbol{a}_k}$ over T for each $k = 1, \ldots, M$ to obtain the derivatives with respect to parameters in the kth layer:

$$\frac{\partial L}{\partial \boldsymbol{a}_k} = \frac{1}{|T|}\sum_{s\in T} \frac{\partial \ell}{\partial \boldsymbol{a}_k}. \tag{9.85}$$

5. Combine all the μ_k-dimensional vectors $\frac{\partial L}{\partial \boldsymbol{a}_k}$, $k = 1, \ldots, M$, into a single μ-dimensional vector:

$$\nabla L(\boldsymbol{a}) = \frac{dL}{d\boldsymbol{a}} = \left(\frac{\partial L}{\partial \boldsymbol{a}_1} \,\middle|\, \frac{\partial L}{\partial \boldsymbol{a}_2} \,\middle|\, \cdots \,\middle|\, \frac{\partial L}{\partial \boldsymbol{a}_M}\right). \tag{9.86}$$

Finally, we can compute the computational complexity of the above steps to obtain the total complexity of the backpropagation algorithm.

0. For each k, computing $A_k h_{k-1}$ requires $O(n_k n_{k_1})$ operations (see (9.15)). Adding b_k and evaluating the activation function $\bar{\lambda}$ both require $O(n_k)$ operations. Therefore, the computational complexity of this step is

$$\sum_{k=1}^{M} O(n_k n_{k-1}) + O(n_k) = \sum_{k=1}^{M} O(n_k n_{k-1} + n_k) = \sum_{k=1}^{M} O(\mu_k) = O(\mu). \tag{9.87}$$

1. As observed above, calculating $\frac{\partial h_{k+1}}{\partial h_k}$ requires $O(n_k n_{k+1})$ operations for each $k < M$. Furthermore, by (9.15), the matrix multiplication (9.80) also requires $O(n_k n_{k+1})$ operations. Finally, we observed above that the calculation of $\partial \ell/\partial h_M$ requires $O(n_M)$ operations. Therefore the total computational complexity of this step is

$$O(n_M) + \sum_{k=1}^{M-1} \mu(n_k n_{k+1}) = \sum_{k=1}^{M-1} O(\mu_{k+1}) = O(\mu). \tag{9.88}$$

2. The multiplication of sparse matrices in (9.82) requires only $O(n_k(n_{k-1}+1)) = O(\mu_k)$ operations since each of the two factor matrices has only one non-zero entry in each column. Therefore, the computational complexity of this step is

$$\sum_{k=1}^{M} O(\mu_k) = O(\mu). \tag{9.89}$$

3. For each k, the matrix $\frac{\partial \ell}{\partial a_k}$ only has one non-zero entry in each column as a result of step 3. Therefore, the matrix multiplication (9.84) requires only $O(\mu_k)$ operations, and the total complexity for this step is

$$\sum_{k=1}^{M} O(\mu_k) = O(\mu). \tag{9.90}$$

Thus, the total computational complexity of the backpropagation algorithm for one object $s \in T$ is $O(\mu)$. In step 4, we calculate $\partial L/\partial \boldsymbol{a}_k$ for each k by repeating steps 0–3 for each $s \in T$. Finally, in step 5, these $\partial L/\partial \boldsymbol{a}_k \in \mathbb{R}^{\mu_k}$ are assembled into a single vector $\partial L/\partial \boldsymbol{a} \in \mathbb{R}^{\mu}$. Because $|T|$ does not depend on μ, repeating steps 0–3 $|T|$ times has $|T|$ times the computational complexity. Furthermore, assembling $\partial L/\partial \boldsymbol{a} \in \mathbb{R}^{\mu}$ requires no additional operations. Thus, the full backpropagation algorithm requires $O(|T|\,\mu)$ operations.

The key step in backpropagation is the backward recursive step (9.80). If we calculate each derivative $\partial \ell/\partial h_k$ independently without this recursive step, then we would replace (9.80) with

$$\frac{\partial \ell}{\partial h_k} = \frac{\partial \ell}{\partial h_M} \prod_{j=k}^{M-1} \frac{\partial h_{j+1}}{\partial h_j}. \tag{9.91}$$

For fixed k, calculating $\partial \ell/\partial h_k$ requires

$$\sum_{j=k+1}^{M} O(n_{j+1} n_j) = \sum_{j=k+1}^{M} O(\mu_j). \tag{9.92}$$

Now repeating the calculation of (9.91) for each $k = 1, \ldots, M-1$ requires

$$\sum_{k=1}^{M} \sum_{j=k+1}^{M} O(\mu_j) = \sum_{k=2}^{M} (k-1) O(\mu_k) = M O(\mu) = O(M\mu) > O(\mu) \tag{9.93}$$

operations, considerably more than backpropagation.

9.6 Exercises

Exercise 1 (Basic computational complexity). Assume that the computational cost of additions, multiplications, and exponentiations are comparable. Give the number of operations necessary to calculate $f(x)$ with

$$f(x) = e^x + 6\,e^{3x} - 5,$$

counting an exponentiation as one operation.

Exercise 2 (Computational complexity and order of operations). Assume that we are given a function $f : \mathbb{R} \to \mathbb{R}$ that is costly to calculate with respect to additions or multiplications.

a. Consider

$$g(x) = f(x) + 2\,(3 + f(x)).$$

Show two different manners of calculating g, one which involves calculating f only once and one which involves calculating f twice.

b. Consider now the more complex formula

$$g(x) = \sum_{k=-5}^{5} (2 + k) f(k^2\,x).$$

i. Explain why for a straightforward calculation of $g(x)$, f must be calculated 11 times.

ii. Show how it is possible to calculate $g(x)$ while calculating f only six times.

Exercise 3 (Computational complexity of the multiplication between a sparse matrix and a full vector). Consider a matrix $M \in M_{N\times N}(\mathbb{R})$. The matrix M has n non-zero entries with $n \ll N$. Consider a vector $u \in \mathbb{R}^N$ with all non-zero entries.

a. Show that the total number of multiplications to calculate $M\,u$ in a straightforward manner is always n.

b. Give two examples of $N \times N$ matrices M_1 and M_2, both with $n \ll N$ non-zero entries, but such that the number of additions needed to calculate $M_1\,u$ is different from the number of additions needed to calculate $M_2\,u$.

Exercise 4 (Another example of the chain rule). We define the function of three variables $\sigma : \mathbb{R}^3 \to \mathbb{R}$

$$\sigma(x, y, z) = \cos(x + 2\,\sin(z + \exp(-y))).$$

Explain how to efficiently calculate all three partial derivatives of σ, following the steps elaborated in Section 9.3.

Exercise 5 (An example of indifferent order of differentiating). We define the function of four variables $\sigma : \mathbb{R}^3 \to \mathbb{R}$

$$\sigma(x, w, y, z) = w^2 + 2x^4 + 5y + 3z^3.$$

Explain why the order in which we calculate the partial derivatives of σ does not affect the computational cost.

Exercise 6 (An example of backpropagation with two neurons per layer).
We consider here how backpropagation works on a simple network with two neuron functions per layer and a total of M layer functions.

a. Explain why the network can be described by a sequence of layer functions h_i with $h_i : \mathbb{R}^2 \to \mathbb{R}^2$ such that our full classifier is a function $h : \mathbb{R}^2 \to \mathbb{R}^2$ with

$$h(x) = h_M \circ \cdots \circ h_1(x).$$

b. We use fully connected layers without bias for each h_i. Explain why that means that each h_i depends on four parameters $a_{i,1}, \ldots, a_{i,4}$ that form a matrix

$$A_i = \begin{bmatrix} a_{i,1} & a_{i,2} \\ a_{i,3} & a_{i,4} \end{bmatrix},$$

so that the function h_i is given by

$$h_i(x) = \bar{\lambda}(A_i\, x),$$

for some non-linear fixed activation function $\bar{\lambda}$.

c. Because of the parameters, we now denote $h_i = h_i(x, a_{i,1}, \ldots, a_{i,4})$ by explicitly including the parameters. Give the formula for the various partial derivatives,

$$\frac{\partial h_i}{\partial x}, \frac{\partial h_i}{\partial a_{i,1}}, \frac{\partial h_i}{\partial a_{i,2}}, \frac{\partial h_i}{\partial a_{i,3}}, \frac{\partial h_i}{\partial a_{i,4}},$$

and explain the computational cost to calculate each of them.

d. Explain how one may use the chain rule to calculate any partial derivative $\partial h/\partial a_{i,j}$ for any $i = 1 \ldots M$ and $j = 1 \ldots 4$.

e. Calculate the number of additions and multiplications required to calculate one partial derivative $\partial h/\partial a_{i,j}$ in terms of i.

f. Explain how to use backpropagation to efficiently calculate all partial derivatives $\partial h/\partial a_{i,j}$ for all $i = 1 \ldots M$ and $j = 1 \ldots 4$. Estimate the number of additions and multiplications that is required.

Exercise 7 (Estimating the cost of forward and backpropagation). We consider for this exercise a straightforward example of a fully linear deep learning neural network. It can be represented by some function $f : \mathbb{R}^N \to \mathbb{R}^N$ which corresponds to K fully linear layers. Each layer k is represented by a function $f_k : \mathbb{R}^N \to \mathbb{R}^N$ given by

$$f_k(x) = \phi(A_k\, x),$$

where A_k is an $N \times N$ matrix and ϕ is for example the absolute value function, $\phi : \mathbb{R}^N \to \mathbb{R}^N$ with

$$(\phi(x))_i = |x_i|.$$

In this context we simply have

$$f(x) = f_K(f_{K-1}(\ldots f_1(x)\ldots)).$$

Finally, we choose as a loss function the sum

$$L : \mathbb{R}^N \to \mathbb{R}, \quad L(x) = \sum_{i=1}^{N} x_i.$$

a. Give the total number of multiplications and additions (counted together) involved in calculating any $f_k(x)$. Do not count taking the absolute values.
b. Deduce the total number of multiplications and additions (again counted together) to calculate $L(f(x))$.
c. Denote by $f_{k,i}$ all coordinates of $f_k(x) = (f_{k,1}(x), \ldots, f_{k,N}(x))$. Estimate the number of operations required to calculate all partial derivatives $\partial f_{k,i}/\partial x_j$.
d. Denote all coefficients in the matrix A_k by $A_{k,mn}$ for $i, j = 1 \ldots N$. Consider now the function f_k as a function of both x and $A_{k,mn}$: $f_k = f_k(x, A_k) = f_k(x, (A_{k,mn})_{mn}) = f_k(x, A_{k,11}, A_{k,12}, \ldots, A_{k,NN})$. Estimate the number of operations required to calculate one partial derivative $\partial f_{k,i}/\partial A_{k,mn}$ for one choice of $m, n \le N$.
e. Estimate the number of operations required to calculate all partial derivatives $\partial f_{k,i}/\partial A_{k,mn}$ for one choice of $m, n \le N$.
f. Consider similarly the function f as a function of x and all $A_{k,mn}$ for all k and m, n: $f = f(x, (A_k)_{k \le K}) = f(x, (A_{k,mn})_{k \le K,\, m,n \le N}) = f(x, A_{1,11}, \ldots, A_{K,NN})$. Reproduce the analysis of Section 9.5 to estimate the total number of operations required to calculate all partial derivatives $\partial L(f)/\partial A_{k,mn}$ for all $k \le K$ and $m, n \le N$.

Exercise 8 (Implementing backpropagation). In this exercise, you will implement backpropagation and observe how much faster it is than finite difference. This is a programming exercise based on Python.

a. Complete the following steps to build a simple DNN in Python without the aid of Pytorch. This DNN has many layers with only one neuron in each layer.
   i. Copy and paste the following code to set the number of layers:

```
num_layers = 30
```

   ii. Complete the following code for the so-called "leaky ReLU function" by replacing ... with commands which return x if $x > 0$ and $0.5x$ otherwise:

```
def ReLU(x):
    ...
```

We will also find it useful to have a function for the derivative of ReLU. Complete the following code for this derivative by replacing ... with commands that return 1 if $x > 0$ and 0.5 otherwise:

```
def dReLU(x):
    ...
```

iii. Copy and paste the following code:

```
def DNN(x, a, b, return_list):
    h = [x]
    for i in range(num_layers):
        h.append(ReLU(a[i]*h[i]+b[i]))
    if return_list:
        return h
    return h[num_layers]
```

This code defines a function DNN accepting four inputs: x is the scalar input to the DNN, a and b are the weight and bias vectors of the DNN, respectively, and return_list is a true/false variable which determines whether the function should return only the output of the last neuron or a list of the output of all neurons (the later option is useful in backpropagation). Then define a function L which accepts a single variable and returns the square of that variable. This will be the loss function.

iv. Copy and paste the following code implementing the backpropagation algorithm which calculates the gradient of the loss function with respect to the parameter vectors a and b:

```
def BP(x, a, b):
    h = DNN(x,a,b,True)
    dLdh = [2*h[num_layers]]
    for i in range(num_layers-1,0,-1):
        dLdh.insert(0,dLdh[0]*dReLU(a[i]*h[i]+b[i])*a[i])
        dLda = [dLdh[i]* dReLU(a[i]*h[i]+b[i])*h[i]
                for i in range(num_layers)]
        dLdb = [dLdh[i]* dReLU(a[i]*h[i]+b[i])
                for i in range(num_layers)]
    return dLda + dLdb
```

Write a short comment after each line describing what it does.

v. Copy and paste the following code which implements the finite difference algorithm which approximates the gradient of the loss function with respect to the parameter vectors a and b:

```
def FD(x, a, b): # Definition of the Finite Difference function
   base = L(DNN(x,a,b,False))
   eps = 0.001 # define a small parameter
   dLda=[0 for i in range(num_layers)]
   dLdb=[0 for i in range(num_layers)]
   for i in range(num_layers): # for-loop over all layers
      aeps = a.copy() # create a copy of the list of parameters a
      aeps[i]+= eps
      beps = b.copy()
      beps[i]+= eps
      dLda[i] = (L(DNN(x,aeps,b,False))-base)/eps
      dLdb[i] = (L(DNN(x,a,beps,False))-base)/eps
   return dLda + dLdb # concatenate two lists
```

Write a short comment after each line of the code which does not have a comment describing what it does.

b. Generate two random lists of parameters a and b of length num_layers as well as a random number x. These values are, respectively, the lists of weights, biases, and the input for the DNN. Copy and paste the following code, which computes the average time taken to compute the gradient of the loss function via finite difference and backpropagation:

```
import time

t1 = time.time()
for i in range(100):
      deriv=BP(my_x,a,b)
t2 = time.time()
for i in range(100):
     deriv = FD(my_x,a,b)
t3 = time.time()


print('avg time for backpropagation: '+str((t2-t1)/100))
print('avg time for finite difference: '+str((t3-t2)/100))
```

What are the times output by this code? Which method is more time consuming, finite difference or backpropagation?

c. Double the variable num_layers and run all the code again. What is the ratio between the computation times obtained in this step and the computation times obtained in part b? Explain how these ratios agree with the computational complexities of backpropagation and gradient descent.

10 Convolutional neural networks

We now introduce *convolutional neural networks* (CNNs), which are a type of deep neural network (DNN) that is well suited for image and audio data when not all, but some local features of the data are important.

CNNs were inspired by the visual perception systems of mammals [46], through the first experiments on the brain's mechanism of vision by Nobel Prize winners Hubel and Wiesel (see [20, 21], and the appendix of [17]).

However, in the earlier implementations of CNNs as in [33] or [13], the backpropagation algorithm had not yet been implemented. In 1989, LeCun et al. applied a new architecture for neural networks to recognize handwritten digits on the MNIST dataset consisting of 28×28 pixel images [28]. While we focus on the simpler architecture of LeCun et al., we do want to emphasize that many recent contributions have dramatically improved upon the CNN architecture we describe here; see for example some of the recent winners [26, 51] of the crowdsourced image labeling competition ILSVRC.

CNNs involve a type of layer naturally called *convolutional layers*, in addition to the fully connected layers that we have studied thus far.

Convolutional layers try to take advantage of the *locality* of features in the data. For example, in facial recognition problems, some of the main features of interest are the eyes, hair, mouth, etc. These features are represented by pixels in the same part of the image, which explains the term locality. In contrast, each neuron in a fully connected layer depends on every pixel in the input image, leading to dependence on non-local information and many more parameters. Because each neuron in a convolutional layer only depends on a small number of pixels in the image, one can hope that the neuron will learn the features in these pixels much more quickly than if it depended on all of the pixels.

CNNs may also use other types of layers that we have not described so far. A common example consists of so-called pooling layers. Each neuron in a max pooling layer compares the values of several neurons in the previous layer and keeps the largest. This helps filtering out less important information in the network. Pooling layers are also helpful because while convolutional layers tend to have fewer parameters than fully connected layers, they generally have many more neurons. Placing a pooling layer following a convolutional layer effectively reduces this number, keeping the total number of neurons in the network manageable.

10.1 Convolution

10.1.1 Convolution of functions

Convolutional layers implement a procedure called *convolution of matrices*, which is closely related to the usual continuous operation of *convolution* on functions defined as follows.

https://doi.org/10.1515/9783111025551-010

Definition 10.1.1. Let, $f, g : \mathbb{R}^n \to \mathbb{R}$ be continuous functions with compact support. Then the *convolution* of f and g is

$$(f * g)(x) = \int_{\mathbb{R}^n} f(y)g(x-y)\,dy. \tag{10.1}$$

Example 10.1.1. Consider the convolution defined in (10.1.1) in one dimension. Choose $g = \chi_{[0,\varepsilon)}$, the *characteristic function* on $[0, \varepsilon)$ defined as

$$\chi_{[0,\varepsilon)} = \begin{cases} 1 & \text{if } x \in [0,\varepsilon), \\ 0 & \text{otherwise.} \end{cases} \tag{10.2}$$

Then consider the convolution of $\frac{1}{\varepsilon}\chi_{[0,\varepsilon)}$ with a continuous function $f : \mathbb{R} \to \mathbb{R}$:

$$\left(f * \frac{1}{\varepsilon}\chi_{[0,\varepsilon)}\right)(x) = \frac{1}{\varepsilon}\int_{\mathbb{R}} \chi_{[0,\varepsilon)}(y)f(x-y)\,dy. \tag{10.3}$$

Since $\frac{1}{\varepsilon}\chi_{[0,\varepsilon)}(y) = 0$ for $y \notin [0,\varepsilon)$, we rewrite (10.3) as

$$\frac{1}{\varepsilon}\int_{\mathbb{R}} \chi_{[0,\varepsilon)}(x)f(x-y)\,dy = \frac{1}{\varepsilon}\int_{0}^{\varepsilon} f(x-y)\,dy. \tag{10.4}$$

Making the change of variables $t = x - y$, we obtain

$$\frac{1}{\varepsilon}\int_{x}^{x+\varepsilon} f(t)\,dt. \tag{10.5}$$

This is the average of f on the interval $(x, x+\varepsilon)$. Therefore, for each x, the function $\frac{1}{\varepsilon}\chi_{[0,\varepsilon)}$ selects a small interval $[x, x+\varepsilon)$ such that the value of the convolution $(\frac{1}{\varepsilon}\chi_{[0,\varepsilon)} * f)(x)$ only depends on $f(t)$ for t in this small interval. For a different function $\bar{f}(t)$ so that $\bar{f}(t) = f(t)$ for $t \in [x, x+\varepsilon)$, the convolution of $\chi_{[0,\varepsilon)}$ with f and $\bar{f}$ is the same at the point x, even if $f(t) \neq \bar{f}(t)$, for $t \in [x, x+\varepsilon)$.

Remark 10.1.1. The convolution operation has many applications in both mathematics and physics, which we do not describe here.

10.1.2 Convolution of matrices

2D grayscale images can naturally be represented by matrices, whose entries encode the pixel intensity values. Some local features can be described by the convolution of matrices, which is the discrete analog of Definition 10.1.2.

To describe this it is useful to rewrite Definition 10.1.1 for $\mathbb{R}^2$ to emphasize the role played by each variable.

Definition 10.1.2. Let $f, g : \mathbb{R}^2 \to \mathbb{R}$ be continuous functions with compact support. Then the *convolution* of f and g is

$$(f * g)(x, y) = \int_{\mathbb{R}^2} f(x', y') g(x - x', y - y')\, dx'\, dy'. \tag{10.6}$$

This inspires the following definition.

Definition 10.1.3 (Convolution of matrices). Let K be a $D \times D$ matrix and let A be an $N \times M$ matrix with $D < N, M$. Then the convolution of K and A is defined as the $(N - D + 1) \times (M - D + 1)$ matrix $B = (b_{i,j})$ such that

$$b_{i,j} = \sum_{n,m=1}^{D} k_{n,m} a_{i+n-1,j+m-1}, \quad \begin{array}{l} i = 1, \dots, N - D + 1, \\ j = 1, \dots, M - D + 1. \end{array} \tag{10.7}$$

We say that B is K convolved with A (indices m and n correspond to the dummy variables in the integration (10.6)).

Remark 10.1.2. The convolution B is smaller than A; A has dimension $N \times M$, but B has dimension $(N - D + 1) \times (M - D + 1)$.

Remark 10.1.3. In (10.7), the matrix A is evaluated at indices $i + n - 1$ and $j + m - 1$, whereas in (10.6), the function g is evaluated at $(x - x', y - y')$. The -1 in both indices is also for technical convenience so that the indices i, j of B start at 1.

Remark 10.1.4. We choose addition in the indices in Definition 10.1.3 (rather than subtraction, which is used in Definition 10.1.2), so that it results in the so-called Hadamard product between matrices as described later (Definition 10.1.4). Often, using the Hadamard product is more convenient because it simplifies notation.

Example 10.1.2 (Simple low-dimension matrix convolution). Consider the two square matrices

$$A = \begin{pmatrix} 0 & 1 & 0 & 1 & 0 \\ 1 & 2 & 1 & 2 & 1 \\ 0 & 1 & 0 & 1 & 0 \\ 1 & 2 & 1 & 2 & 1 \\ 0 & 1 & 0 & 1 & 0 \end{pmatrix} \tag{10.8}$$

and

$$K = \begin{pmatrix} 0 & 1 & 0 \\ 1 & 2 & 1 \\ 0 & 1 & 0 \end{pmatrix}. \tag{10.9}$$

Calculate the matrix $K * A$. Since A is 5×5 and K is 3×3, we have $N = M = 5$ and $D = 3$. Then the size of $K * A$ is $(N - D + 1) \times (M - D + 1)$, or 3×3. We will use (10.7) to calculate each entry of $K * A(\alpha, \beta)$, for $1 \leq \alpha, \beta \leq 3$. Due to the symmetries in A, many of the entries of $K * A$ are the same. In fact, there are only three distinct values of the entries of $K * A$, which we show explicitly for $(1, 1)$:

$$\begin{aligned}
(K * A)(1,1) &= (K * A)(3,1) = (K * A)(1,3) = (K * A)(3,3) \\
&= \sum_{\alpha',\beta'=0}^{3} K(\alpha',\beta')A(1+\alpha'-1, 1+\beta'-1) \\
&= K(1,1)A(1,1) + K(1,2)A(1,2) + K(1,3)A(1,3) \\
&\quad + K(2,1)A(2,1) + K(2,2)A(2,2) + K(2,3)A(2,3) \\
&\quad + K(3,1)A(3,1) + K(3,2)A(3,2) + K(3,3)A(3,3) \\
&= 0\cdot 0 + 1\cdot 1 + 0\cdot 0 + 1\cdot 1 + 2\cdot 2 + 1\cdot 1 + 0\cdot 0 + 1\cdot 1 + 0\cdot 0 \\
&= 8. \qquad (10.10)
\end{aligned}$$

The other entries are calculated similarly, so we have

$$B := K * A = \begin{pmatrix} 8 & 6 & 8 \\ 6 & 4 & 6 \\ 8 & 6 & 8 \end{pmatrix}. \qquad (10.11)$$

10.1.3 Hadamard product and feature detection

Convolution can also be understood via an inner product on matrices called the Hadamard product, defined as follows.

Definition 10.1.4 (Hadamard product of matrices). Let A, B be $D \times D$ matrices. Then the Hadamard product of A and B, denoted $A \otimes B$, is

$$A \odot B = \sum_{i,j=1}^{D} a_{i,j} b_{i,j} \in \mathbb{R}. \qquad (10.12)$$

We will see that the convolution of a $D \times D$ kernel K and a matrix A is equal to a matrix of Hadamard products between K and the various $D \times D$ *submatrices* of A, which are defined as follows.

Definition 10.1.5 (Submatrix of a matrix). Let $A = (a_{i,j})$ be an $N \times M$ matrix and let $D > 0$. The $D \times D$ submatrix of A at indices (i, j) with $1 \leq i \leq N - D + 1$ and $1 \leq j \leq M - D + 1$ is

$$A[i, j; D] = \begin{pmatrix} a_{i,j} & \cdots & a_{i,j+D-1} \\ \vdots & \ddots & \vdots \\ a_{i+D-1,j} & \cdots & a_{i+D-1,j+D-1} \end{pmatrix}. \qquad (10.13)$$

Simply put, $A[i,j;D]$, the $D \times D$ submatrix of A at indices i, j, consists of the entries of A in rows i through $i + D - 1$ and columns j through $j + D - 1$.

The Hadamard product and the notion of submatrices leads to the following equivalent definition of convolution.

Definition 10.1.6 (Definition of matrix convolution as Hadamard product). Let K be a $D \times D$ kernel matrix and let A be an $N \times M$ matrix with $D < \min\{N, M\}$. Then the convolution matrix $B = K * A$ is defined by its entries $b_{i,j}$:

$$b_{i,j} = K \odot B[i,j;D], \qquad \begin{array}{l} i = 1, \dots N - D + 1, \\ j = 1, \dots M - D + 1. \end{array} \tag{10.14}$$

Formula (10.14) shows that the (i, j) entry of the convolution matrix B is the Hadamard product of K and the submatrix of A of the same size at indices i, j (its top left entry is $a_{i,j}$).

Example 10.1.3 (Feature detection via the Hadamard product). Consider the matrices A and K from Example 10.1.2 defined by (10.8) and (10.9), respectively. The entry $b_{1,1} = 8$ in (10.11) is computed as the Hadamard product of the upper left 3×3 submatrix $A[1,1;3]$ and K:

$$A[1,1;3] = \begin{pmatrix} 0 & 1 & 0 \\ 1 & 2 & 1 \\ 0 & 1 & 0 \end{pmatrix}, \quad K = \begin{pmatrix} 0 & 1 & 0 \\ 1 & 2 & 1 \\ 0 & 1 & 0 \end{pmatrix}. \tag{10.15}$$

Since the two matrices happen to be the same, the Hadamard product leads to $b_{1,1}$ being the largest value among all the entries of B. Indeed, if we consider all entries of the matrix as a vector, then (10.12) becomes a standard inner product. We next observe that each element of the convolution (10.11) is a Hadamard product of a 3×3 submatrix of A and the 3×3 matrix K. Recall the formula for the inner product of $x \cdot y = |x||y| \cos(\gamma)$ and therefore is largest when the vectors are parallel, i. e., $\cos(\gamma) = 1$, or when $x = \lambda y$ for any scalar $\lambda > 0$.

This can be interpreted as follows. The matrix $A[1,1;3]$ represents a feature in an image that is of primary interest to us. Then we choose the kernel K which mimics this feature as closely as we can; in our case we choose $K = A[1,1;3]$. Now a computer can detect the feature by finding the largest values in the convolution matrix B instead of searching for features represented by submatrices in the original image A. This demonstrates how convolution reduces the problem of feature detection to a simple problem of finding the largest entries in a matrix by choosing a convolution kernel which best mimics the feature.

Example 10.1.4 (Vertical edge detection via convolution). Consider a grayscale image described by a matrix whose entries are in $\{0, 1\}$:

$$A = \begin{pmatrix} 0 & 0 & 1 & 1 & 1 \\ 0 & 0 & 1 & 1 & 1 \\ 0 & 0 & 1 & 1 & 1 \\ 0 & 0 & 1 & 1 & 1 \\ 0 & 0 & 1 & 1 & 1 \end{pmatrix}. \tag{10.16}$$

The corresponding image is white on the left half and black on the right half, with the two regions separated by a vertical edge in the center (Fig. 10.1). In order to detect the edge in the image from the matrix A directly, a computer would have to compare each pair of horizontally adjacent entries of A to find where there is a jump from 0 to 1 corresponding to a jump in color (from white to black). Instead, we can consider the convolution with the following kernel K:

$$K = \begin{pmatrix} -1 & 1 \\ -1 & 1 \end{pmatrix}. \tag{10.17}$$

Then the resulting convolution matrix B is

$$B := K * A = \begin{pmatrix} 0 & 2 & 0 & 0 \\ 0 & 2 & 0 & 0 \\ 0 & 2 & 0 & 0 \\ 0 & 2 & 0 & 0 \end{pmatrix}. \tag{10.18}$$

The kernel K is chosen so that 2 appears as a jump between two horizontally adjacent matrix entries or the pixels corresponding to these entries. Therefore, the column of 2s traces the vertical edge.

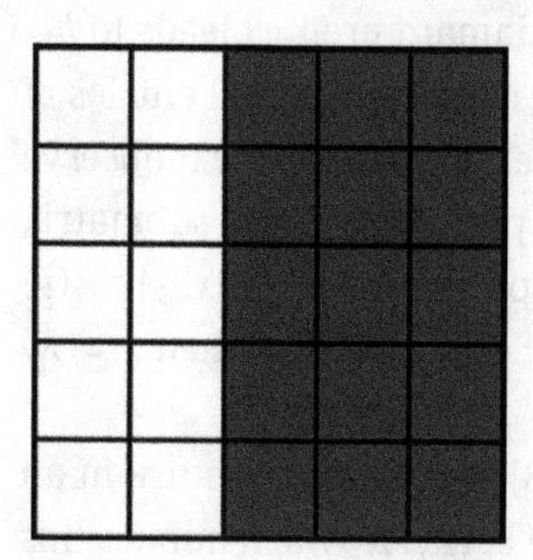

Figure 10.1: Each square in the grid represents a pixel that corresponds to an entry in matrix (10.16). White pixels are represented by 0s and black pixels are represented by 1s.

Note that the values of −1 and 1 in each row of K result in the convolution detecting the difference between horizontally adjacent pixels. Indeed, convolution of K with the 2×2 submatrix of the image matrix A,

$$\bar{A} = \begin{pmatrix} a & b \\ c & d \end{pmatrix}, \tag{10.19}$$

gives the following explicit formula for the (i, j) entry of the convolution matrix:

$$(K * \bar{A})_{ij} = b - a + d - c. \tag{10.20}$$

The two cases for the relative values of a, b, c, d in $\bar{A}$ which appear in A are:

a) If entries of $\bar{A}$ are all zero or all one, i. e., $a = b = c = d$, then from (10.20) we get $K * \bar{A} = 0$, meaning there is no edge in $\bar{A}$.
b) If the first column in $\bar{A}$ is all zeros and the second is all ones, i. e., $a < b$ and $c < d$, then from (10.20) we get $K * \bar{A} > 0$, i. e., there is a jump in pixel intensity which identifies an edge in the image.

Remark 10.1.5. We focus for simplicity on the convolution of matrices corresponding to, e. g., 2D images. However, one could define the convolution of tensors of any rank from 1 (for, e. g., audio data) to 2 (for, e. g., images) to 3 (for, e. g., video), and so on.

10.2 Convolutional layers

Applying Definition 4.1.2 of the layer function to the kth layer, we recall that the kth layer function $h_k : \mathbb{R}^{n_{k-1}} \to \mathbb{R}^{n_k}$ between the $(k-1)$th and the kth layer is

$$h_k = \bar{\lambda}(A_k \cdot h_{k-1} + b_k), \tag{10.21}$$

where A_k is the weight matrix of the kth layer, b_k is the bias vector of the kth layer, $\bar{\lambda}$ is an activation function, and h_{k-1} is the output of the previous $(k-1)$th layer.

We next define the convolutional layer function, which is a basic building block of CNNs. Heuristically, convolutional layer functions can be obtained from fully connected layer functions by fixing certain weights to be zero so these weights “cannot be trained,” that is, they are not part of the minimization of the loss function. Moreover, the non-zero weights are shared between neurons, as explained in Example 10.2.1 below.

Definition 10.2.1 (Convolutional layer function). Let K be a $D \times D$ kernel matrix of parameters and let N and M be integers such that $D \leq N, M$. A *convolutional layer function* is a mapping $c : \mathbb{R}^{N \times M} \to \mathbb{R}^{(N-D+1)\times(M-D+1)}$ of the following form:

$$c(X) = \bar{\lambda}(K * X + b), \tag{10.22}$$

where X is an $N \times M$ matrix, b is the usual bias represented here by an $(N-D+1)\times(M-D+1)$ matrix, and $\bar{\lambda}$ is an activation function applied to each entry of the matrix $K * X + b$ defined in Definition 10.1.3. The matrices X and $c(X)$ are called the input and output of the layer function, respectively.

Observe that matrix $c(X)$ is smaller than matrix X; see Remark 10.1.2.

Definition 10.2.2 (Convolutional neuron function). Let K be a $D \times D$ kernel matrix of parameters and let N and M be integers such that $D \leq N, M$. Then for integers i, j such that $1 \leq i \leq N - D + 1$ and $1 \leq j \leq M - D + 1$, a *convolutional neuron function* is a mapping $q_{i,j} : \mathbb{R}^{N \times M} \to \mathbb{R}$ of the form

$$q_{i,j}(X) = K \odot X[i, j; D] + b_{i,j}, \tag{10.23}$$

where X is an $N \times M$ matrix, $X[i, j; D]$ is the submatrix of X defined by (10.13), b is an $(N - D + 1) \times (M - D + 1)$ matrix, and $\odot$ is the Hadamard product defined in (10.12).

Definition 10.1.6 shows that the convolutional layer function $c(X)$ is a matrix whose entries are convolutional neuron functions $q_{i,j}(X)$:

$$c(X) = \begin{pmatrix} q_{1,1}(X) & \cdots & q_{1,M-D+1}(X) \\ \vdots & \ddots & \vdots \\ q_{N-D+1,1}(X) & \cdots & q_{N-D+1,M-D+1}(X) \end{pmatrix}. \tag{10.24}$$

Each entry $q_{i,j}(X)$ is the Hadamard product of K with the submatrix of X of the same size at indices i, j, plus the biases.

In fully connected layers, we usually represent the input and output as vectors (see Definition 4.1.2). In contrast, in the convolutional layer defined above, it is easier to represent the input and output as matrices because of Definition 10.1.3 for matrix convolution. However, one may also represent the pixels as a vector if it is more convenient for a given algorithm.

Remark 10.2.1. In a CNN, the convolutional layer function c in (10.22) is the kth layer function in the network, i. e., $c(X) = c_k(X)$. The kernel $K = K_k$ has size $D_k \times D_k$ and

$$c_k : R^{N_{k-1} \times M_{k-1}} \to \mathbb{R}^{N_k \times M_k}, \tag{10.25}$$

where $N_{k-1} \times M_{k-1}$ is the size of the output of the $(k-1)$th layer (i. e., the number of neurons in the $(k-1)$th layer), $N_k = N_{k-1} - D_k + 1$, and $M_k = M_{k-1} - D_k + 1$. The number of neurons in the kth layer is $N_k \cdot M_k$. The matrix X in (10.22) is the output of the $(k-1)$th layer and the input of the kth layer, whereas $c_k(X)$ is the output of the kth layer. If $k = 1$, then X is the matrix corresponding to the image and c_1 is the first layer.

One advantage of convolutional layers is that for the same size of input data, they use *fewer parameters* (weights) than a fully connected layer. Indeed, the fully connected layer function (10.21) depends on $n_k(n_{k-1} + 1)$ parameters. Since typically $N_k, N_{k-1} \gg 1$, the number of parameters is much larger than both N_k, N_{k-1}. By contrast, the kth convolutional layer function c_k depends only on D_k^2 parameters. Typically, D_k is much less than any of N_k, M_k, N_{k-1}, and M_{k-1}, so the number of parameters in a convolutional layer is much less than either the number of neurons in the layer or the number of neurons in the previous layer.

Example 10.2.1 (Weight sharing and sparsity). In this example, we compare the definition of the layer function (Definition 4.1.2) with the definition of the convolutional layer function (Definition 10.2.1). The idea is to rewrite the definition of a convolutional layer in terms of matrix multiplications rather than convolutions. Consider the following two matrices for a convolution kernel and input to a convolutional layer function, respectively:

$$K := \begin{pmatrix} a & b \\ c & d \end{pmatrix}, \quad X := \begin{pmatrix} 1 & 2 & 3 \\ 4 & 5 & 6 \\ 7 & 8 & 9 \end{pmatrix}. \tag{10.26}$$

Applying Definition 10.1.3, we obtain

$$K * X = \begin{pmatrix} 1 \cdot a + 2 \cdot b + 4 \cdot c + 5 \cdot d & 2 \cdot a + 3 \cdot b + 5 \cdot c + 6 \cdot d \\ 4 \cdot a + 5 \cdot b + 7 \cdot c + 8 \cdot d & 5 \cdot a + 6 \cdot b + 8 \cdot c + 9 \cdot d \end{pmatrix}. \tag{10.27}$$

Next, rewrite the matrices $X \in \mathbb{R}^{3\times3}$ and $K * X \in \mathbb{R}^{2\times2}$ as a 9D vector $\bar{X}$ and a 4D vector $\overline{K * X}$, respectively:

$$\bar{X} = (1 \quad 2 \quad 3 \quad 4 \quad 5 \quad 6 \quad 7 \quad 8 \quad 9)^T, \tag{10.28}$$

$$\overline{K * X} = \begin{pmatrix} 1 \cdot a + 2 \cdot b + 4 \cdot c + 5 \cdot d \\ 2 \cdot a + 3 \cdot b + 5 \cdot c + 6 \cdot d \\ 4 \cdot a + 5 \cdot b + 7 \cdot c + 8 \cdot d \\ 5 \cdot a + 6 \cdot b + 8 \cdot c + 9 \cdot d \end{pmatrix}. \tag{10.29}$$

These vectors are related via the usual matrix-vector multiplication:

$$\overline{K * X} = \mathcal{K}\bar{X}, \tag{10.30}$$

where $\mathcal{K}$ is the matrix

$$\mathcal{K} = \begin{pmatrix} a & b & 0 & c & d & 0 & 0 & 0 & 0 \\ 0 & a & b & 0 & c & d & 0 & 0 & 0 \\ 0 & 0 & 0 & a & b & 0 & c & d & 0 \\ 0 & 0 & 0 & 0 & a & b & 0 & c & d \end{pmatrix}. \tag{10.31}$$

Using (10.30), a convolutional layer $c(X) = \bar{\lambda}(K * X)$ with kernel K can be rewritten as a fully connected layer $h(\bar{X}) = \bar{\lambda}(\mathcal{K}\bar{X})$ with parameter matrix $\mathcal{K}$. The layer function h has four neuron functions corresponding to the four rows of $\mathcal{K}$. Observe that while in general the weight matrix of a fully connected layer may have all elements non-zero and distinct, the weight matrix $\mathcal{K}$ corresponding to the kernel K has mostly 0 entries and only 15 non-zero entries corresponding to the four distinct numbers a, b, c, d. For example, in Fig. 10.2 the dashed connections on the right (in the fully connected layer) that are not present on the left (in the convolutional layer) are assigned 0 weights. In other words, the convolutional layer can be represented by a fully connected layer with a sparse matrix of parameters.

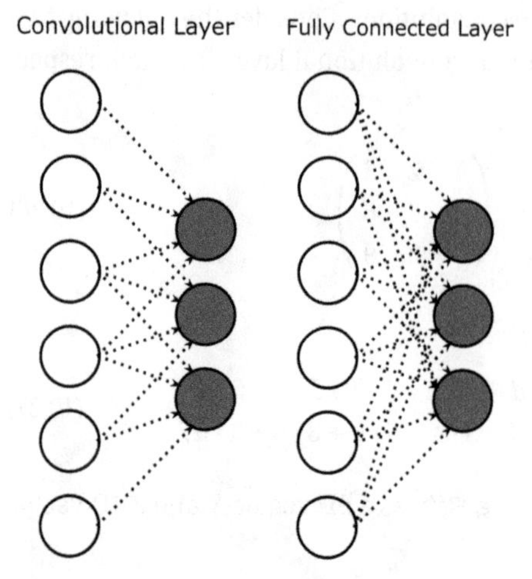

Figure 10.2: *Left*: Convolutional layer with the output (gray) neurons depending on four nodes of the previous layer denoted by dashed arrows from the previous layer to the next layer. Equivalently, this convolutional layer uses a 2×2 kernel K and thus contains four weights. These four weights are assigned to the four dashed lines entering each gray neuron. The number of weights does not depend on the number of neurons in either the previous or the current layer; it only depends on the size of the kernel. *Right*: Fully connected layer where every node in the output depends on every node in the input. The number of weights is the product of the number of neurons in input and output layers, i. e., $6 \times 3 = 18$. In contrast to the convolutional layer, the number of weights increases as the number of nodes in either layer increases.

In fact, each of the four neurons of a fully connected layer h (i. e., four rows of $\mathcal{K}$) have the same four non-zero weights a, b, c, and d (though not all in the same columns). In this sense, all neuron functions of a convolutional layer *share* the same set of weights, e. g., a, b, c, and d in the above examples. Furthermore, these weights are entries of the convolution kernel K. This property is called *weight sharing* and it reflects the reduction of parameters when using convolutional layers versus using fully connected layers. Finally we note that just as in a fully connected layer, the weights (kernel) of a convolutional layer are trained via, e. g., gradient descent.

10.3 Padding layer

A common type of layers often employed together with CNNs are so-called padding layers. Padding layers typically add several rows and columns of 0 (or another constant) to an image near its boundary.

There can be several reasons to add such a layer. They are commonly used when the initial dataset is composed of objects that do not all have the same dimensions. For example, our dataset may correspond to a collection of grayscale images with different resolutions: An image may have N horizontal pixels and M vertical pixels, while another image has $\tilde{N}$ horizontal pixels and $\tilde{M}$ vertical pixels with $N \neq \tilde{N}$ and/or $M \neq \tilde{M}$.

This may present a challenge when applying DNN techniques: First of all, numerical implementations can be difficult with inputs of varying sizes. Second, the architecture of a network often presumes a given size for the input. This is particularly obvious if the first layer is fully connected: As a fully connected layer includes a multiplication by a matrix, the sizes would have to match.

In such a setting, padding can be added to each image separately. This does not affect the information contained in each one of the images, but it is a straightforward solution to make them all of identical size.

Padding layers can also be specifically useful when combined with CNNs. Observe that in the definition of the convolution of matrices (see Definition 10.1.3), the entry $a_{1,1}$ is only used once, when calculating $b_{1,1}$, and it is not used when calculating any $b_{i,j}$ with $i > 1$ or $j > 1$. Similarly, one can check that an entry on the first row, $a_{1,j}$ for $1 < j < M$, or first column, $a_{i,1}$ for $1 < i < M$, is used D times. However, any entry in the middle of the matrix, a_{ij} with $D \le i, j \le N - D + 1$, will be used D^2 times in the calculations.

Consequently, whichever type of information or feature was contained in the first or last rows/columns of the matrix a is weighted less on the resulting matrix b than features encoded in the center of a. This means that CNNs may be less sensitive to features near the boundary of images. Padding layers offer a straightforward solution to this issue: by adding rows and columns of 0s around a, we can make it so that what were the first/last rows/columns are now enough in the center to be used as many times as other entries. We will also use padding layers later in this chapter in a somewhat similar manner to construct a DNN that is translationally invariant.

In mathematical terms, given a matrix M in $\mathbb{R}^{n\times n}$, a padding layer with padding size $2s$ will add s components to the matrix at every extremity and fill them with 0, leading to the following definition.

Definition 10.3.1. A padding layer function with padding size $2s$ is a function $P : \mathbb{R}^{n\times n} \to \mathbb{R}^{(n+2s)\times(n+2s)}$ defined by

$$(P(M))_{\alpha,\beta} = M_{\alpha-s,\beta-s}, \quad \text{if } s+1 \le \alpha,\ \beta \le n+s,$$

and

$$(P(M))_{\alpha,\beta} = 0, \quad \text{if } \alpha \le s,\ \alpha \ge n+s+1,\ \beta \le s, \text{ or } \beta \ge n+s+1.$$

10.4 Pooling layer

In this section we will discuss another type of layer that may be found in CNNs: *pooling layers*. Pooling layers may for example alternate with convolutional layers. The input of a pooling layer is the output matrix of the previous convolutional layer. The idea is that pooling layers remove smaller entries and keep large entries of this matrix which hopefully correspond to the more important features. Mathematically, it means that the output of the pooling layer has a smaller size than the output matrix of the previous convolutional layer.

While there are many ways to perform pooling, for simplicity we focus here on *max pooling*. We first introduce the entrywise max function on a matrix $\overline{\max} : \mathbb{R}^{N\times N} \to \mathbb{R}$ which chooses the maximum element of a matrix. This definition will be applied below to submatrices of a larger matrix.

Definition 10.4.1. Let $X \in \mathbb{R}^{N\times N}$ be an $N \times N$ matrix. The entrywise max function $\overline{\max} : \mathbb{R}^{N\times N} \to \mathbb{R}$ is defined as

$$\overline{\max}(X) := \max_{1\le i,j\le N} X_{ij}. \tag{10.32}$$

We next write X as a block matrix whose elements $X[i,j;D]$ are themselves $D \times D$ submatrices as in Definition 10.1.5. Every element of X is in exactly one of the blocks $X[i,j;D]$, that is, the blocks are disjoint. For example, we choose $D = 2$; if

$$X = \begin{pmatrix} x_{11} & x_{12} & \cdots & x_{1N} \\ x_{21} & x_{22} & \cdots & x_{2N} \\ \vdots & \vdots & \ddots & \vdots \\ x_{N1} & x_{N2} & \cdots & x_{NM} \end{pmatrix}, \tag{10.33}$$

then the block $X[1,1;2]$ is the following submatrix of X:

$$X[1,1;2] = \begin{pmatrix} x_{11} & x_{12} \\ x_{21} & x_{22} \end{pmatrix}. \tag{10.34}$$

The max pooling layer is defined by applying the elementwise max function to each block.

Definition 10.4.2 (Max pooling layer function). Let $X \in \mathbb{R}^{N\times M}$ be a matrix represented by $D \times D$ block matrices $X[i,j;D]$, $i = 1,\dots,S = N/D$ and $j = 1,\dots T = M/D$:

$$\begin{pmatrix} X[1,1;D] & X[1,D+1;D] & \cdots & X[1,(T-1)D+1;D] \\ X[D+1,1;D] & \ddots & \cdots & X[D+1,(T-1)D+1;D] \\ \vdots & \vdots & \ddots & \vdots \\ X[(S-1)D+1,1;D] & \cdots & \cdots & X[(S-1)D+1,(T-1)D+1;D] \end{pmatrix}. \tag{10.35}$$

Then the max pooling layer function $p : \mathbb{R}^{N\times M} \to \mathbb{R}^{S\times T}$ is defined as

$$p(X) := \begin{pmatrix} \overline{\max}(X[1,1;D]) & \cdots & \overline{\max}(X[1,(T-1)D+1;D]) \\ \vdots & \ddots & \vdots \\ \overline{\max}(X[(S-1)D+1,1;D]) & \cdots & \overline{\max}(X[(S-1)D+1,(T-1)D+1;D]) \end{pmatrix}. \tag{10.36}$$

Thus, the max pooling layer function reduces the dimension of the original matrix X by replacing each block matrix with the maximum element over that block.

Example 10.4.1 (Edge detection). Consider the convolution matrix $B = K * A$ obtained in (10.18). We represent the matrix B via 2×2 block matrices $\{X^{ij}\}_{i,j=1}^{2}$:

$$B = \begin{pmatrix} X[1,1;2] & X[1,3;2] \\ X[3,1;2] & X[3,3;2] \end{pmatrix} = \left(\begin{array}{cc|cc} 0 & 2 & 0 & 0 \\ 0 & 2 & 0 & 0 \\ \hline 0 & 2 & 0 & 0 \\ 0 & 2 & 0 & 0 \end{array}\right). \tag{10.37}$$

Then we apply max pooling to obtain the following matrix:

$$p(B) = \begin{pmatrix} 2 & 0 \\ 2 & 0 \end{pmatrix}, \tag{10.38}$$

in which only the largest element of each 2×2 submatrix is retained after max pooling.

The pooling layer reduces the complexity of a CNN by removing unimportant data at the cost of losing fine details of the features. For example, the upper left 2×2 block $X[1,1;2]$ in Example 10.4.1 is replaced by one number, representing the feature. In the block matrix, we know that there is at least one edge and we know the location: between two white pixels in the left column and two black pixels in the right column. In (10.38) we only know that there is at least an edge but we cannot easily reconstruct where it should be.

10.5 Building CNNs

A common architecture of CNNs can be described as follows: They typically consist of convolutional layers alternating with pooling layers with sometimes fully connected layers being used before or after. See Fig. 10.3 in 1D and Fig. 10.4 in 2D for a simple CNN with one convolutional layer, pooling layer, and fully connected layer.

For simplicity of presentation, we only consider here CNNs from [29] that consist of sequences of alternating convolutional layers c_i and pooling layers p_i followed by a fully connected layer $h(x)$. Thus, we define a CNN as follows:

$$\phi(X) := h \circ p_K \circ c_K \circ \cdots \circ p_1 \circ c_1(X). \tag{10.39}$$

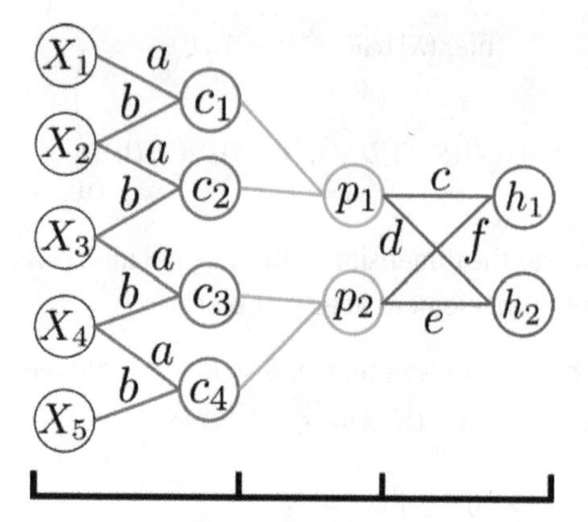

Figure 10.3: A simple CNN with three layers: convolutional (in red on the left), pooling (in green in the middle), and fully connected (in blue on the right).

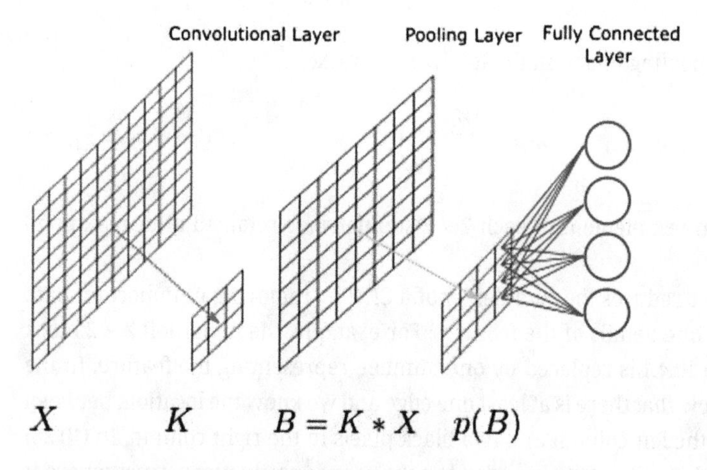

Figure 10.4: Schematic representation of convolutional matrix $c(X)$, kernel matrix K, pooling layer matrix $p(c(X))$, and the fully connected layer vector $h(p(c(X)))$.

In practice, more general CNNs are used; see [26, 51]. In particular, padding layers are used in CNNs.

A basic idea behind this choice of architecture can be summarized as follows. Convolutional layers identify local features in an image via an output matrix whose large entries correspond to these features. Then pooling layers reduce the size of the convolutional output matrices by discarding their small entries. Finally, fully connected layers make classifications based on the local features that are found via convolutional layers and their relative position.

For simplicity of presentation, we considered above only one kernel per convolutional layer. In applications, multiple kernels are used, leading to so-called multichannel convolutional layers. Each kernel corresponds to a different feature which is learned simultaneously. A multichannel convolutional layer may receive input in the form of several matrices (corresponding to several input channels), and provides as output several matrices (corresponding to several output channels).

Indeed, the input data are often multichannel as well. In color images, for example, the color of each pixel may be described by the intensity of red, green, and blue in the pixel (this is called RGB encoding). Thus, a single $N \times M$ image is represented by three matrices: one for the intensity of each color. Each matrix is called a *channel* of the image. Audio data can be multichannel as well, for example, stereo music contains one channel for the left ear and one for the right to provide a directional quality to the sound.

In such a case, the inputs and outputs of convolutional and pooling layers need to be vectorized.

Definition 10.5.1. Let $\{K_{i,j}\}_{i=1\ldots k,\, j=1\ldots\ell}$ be $D_j \times D_j$ matrices of parameters with $k, \ell \geq 1$ positive integers corresponding to the number of input and output channels. A *multichannel convolutional layer function* is a mapping $c : \mathbb{R}^{N\times M\times k} \to \mathbb{R}^{(N-D_1+1)\times(M-D_1+1)} \times \mathbb{R}^{(N-D_2+1)\times(M-D_2+1)} \times \cdots \mathbb{R}^{(N-D_\ell+1)\times(M-D_\ell+1)}$ of the following form:

$$c(X) = \left(\bar{\lambda}\left(\sum_{i=1}^{k} K_{i1} * X_i\right), \bar{\lambda}\left(\sum_{i=1}^{k} K_{i2} * X_i\right), \ldots, \bar{\lambda}\left(\sum_{i=1}^{k} K_{i\ell} * X_i\right)\right), \tag{10.40}$$

where the input is represented by k matrices $(X_i)_{i=1\ldots k}$ of size $N \times M$.

Similarly, a multichannel pooling layer is defined as follows.

Definition 10.5.2. Let $\{X_i\}_{i=1}^{\ell}$ be ℓ matrices of dimensions $N_i \times M_i$ represented by $\tilde{D}_i \times \tilde{D}_i$ block matrices $\{X^{ij}\}$. Then the multichannel pooling layer $\bar{p} : \mathbb{R}^{N_1\times M_1} \times \mathbb{R}^{N_2\times M_2} \times \cdots \mathbb{R}^{N_n\times M_n} \to \mathbb{R}^{S_1\times T_1} \times \mathbb{R}^{S_2\times T_2} \times \cdots \mathbb{R}^{S_n\times T_n}$ where $S_i = N_i/\tilde{D}_i$ and $T_i = M_i/\tilde{D}_i$ for some choice of $\tilde{D}_i$ is defined as

$$\bar{p} = (p(X_1), p(X_2), \ldots, p(X_\ell)), \tag{10.41}$$

where $N_i \times M_i = (N - D_i + 1) \times (M - D_i + 1)$ from Definition 10.5.1 when a pooling layer directly follows a convolutional layer.

10.6 Equivariance and invariance

A difficult but interesting question is what kind of invariance a DNN may possess. For example, the training set may contain images which are related via some transformation, e. g., one image may be obtained by rotating another. Then, the invariance property or simply invariance would mean that these images would be automatically classified into the same class.

On the other hand, a DNN classifier function may not be automatically invariant under rotation for all choices of parameters. But one may wonder if there exists some choice of DNN parameters (through a proper choice of architecture such as the number of layers, their widths, or connectivity of layers) for which the network classifier function is invariant.

Understanding which DNN architecture leads to which type of invariance is hence important but also usually complex to answer. Deep convolutional networks in particular have been recognized to be able to identify progressively more effective and intricate invariance; see for example [2, 7, 27]. We also refer the reader to the recent approach to this question through the notion of scattering transform in [6, 31] and to the review [32]. In this section, we first focus on the question of translational invariance. To this end, we first introduce a more general notion of equivariance. Once translational equivariance is introduced and explained, we conclude with an example of a translationally invariant CNN.

A remarkable feature of CNN classifiers is that they are able to correctly classify translations of images. Several images can have the same local feature in different locations. For example, Fig. 10.5 shows two images, each with the same pen in different locations. Because CNNs do not learn pixel representations of the local feature but rather learn feature representations, they typically classify both images in Fig. 10.5 in the same class.

(a) A pen on the left (b) A pen on the right

Figure 10.5: Two images, both of which contain a pen. In one image the pen is on the left and in the other the pen is on the right. A convolutional layer will identify a pen in both images, unlike fully connected layers, which would see the two images as entirely unrelated objects.

More specifically, convolutional layer functions $c(X)$ (see Definition 10.2.1) possess the property that when a feature (e. g., the pen in Fig. 10.5a) is shifted in the input image X, there is a corresponding shift in the output matrix $c(X)$ such as in Fig. 10.6. This property is called *translational equivariance*, which is described in a simple setting below.

To define translational equivariance for matrices, we first introduce the *horizontal and vertical shift operators* for matrices, defined as follows.

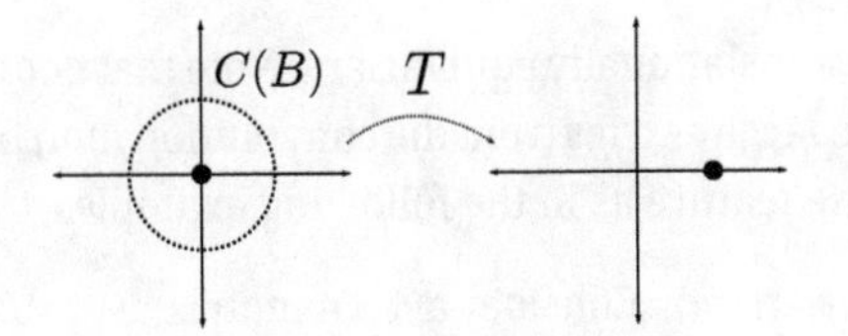

(a) Computing the center C of the ball B and then translating by T

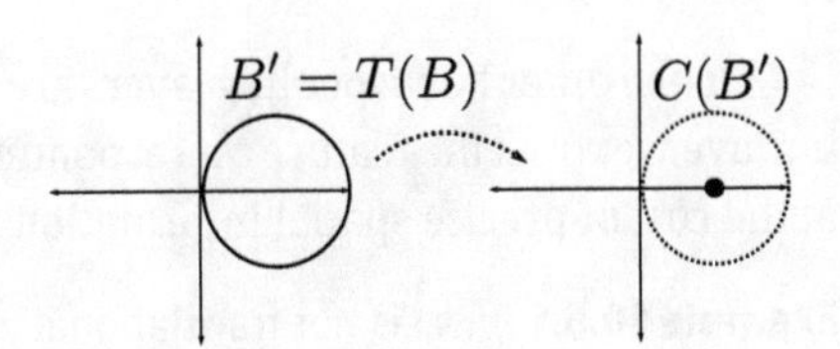

(b) Translating the ball B by T and then computing the center C

Figure 10.6: The center C of a ball B is equivariant under translations T, that is, $C \circ T(B) = T \circ C(B)$ for any ball $B \subset \mathbb{R}^2$.

Definition 10.6.1. Let $\mathbb{R}^{N\times M}$ be the vector space of $N \times M$ matrices. The horizontal shift operator $T_h : \mathbb{R}^{N\times M} \to \mathbb{R}^{N\times M}$ is defined as

$$(T_h(A))_{i,j} = \begin{cases} a_{i,j-1} & 2 \le j \le M, \\ a_i, M & j = 1. \end{cases} \tag{10.42}$$

The vertical shift operator $T_v : \mathbb{R}^{N\times M} \to \mathbb{R}^{N\times M}$ is defined similarly:

$$(T_v(A))_{i,j} = \begin{cases} a_{i-1,j} & 2 \le i \le N, \\ a_N, j & i = 1. \end{cases} \tag{10.43}$$

Simply put, $T_h(A)$ shifts each entry of A horizontally to the next column and moves the entries of A in the last column to the first column. Similarly, $T_v(A)$ shifts each entry of A vertically to the next row and moves the entries of the last row to the first row. Then translational equivariance is defined via the translation operators a follows.

Definition 10.6.2. Let $\mathbb{R}^{N_1\times M_1}$ and $\mathbb{R}^{N_2\times M_2}$ be the spaces of $N_1 \times M_1$ and $N_2 \times M_2$ matrices, respectively. Let $f : \mathbb{R}^{N_1\times M_1} \to \mathbb{R}^{N_2\times M_2}$ be a function; f is translationally equivariant if

$$f \circ T_h(A) = T_h \circ f(A) \quad \text{and} \quad f \circ T_v(A) = T_v \circ f(A) \tag{10.44}$$

for each $A \in \mathbb{R}^{N_1\times M_1}$.

We see from Definition 10.6.2 that f is translationally equivariant if and only if it commutes with the vertical and horizontal shift operators. This is because any translation can be decomposed into a composition of horizontal and vertical shifts.

For a given kernel K, the convolutional layer function $c(X)$ as defined in Definition 10.2.1 is not translationally equivariant, but it is *almost* translationally equivariant in the following sense: For indices $i, j \ge 2$,

$$(c \circ T_h(X))_{ij} = (T_h \circ c(X))_{ij} \quad \text{and} \quad (c \circ T_v(X))_{ij} = (T_v \circ c(X))_{ij}, \tag{10.45}$$

that is, the convolutional layer commutes with the vertical and horizontal shift operators in all of its entries except for those in the first row ($i = 1$) and first column ($j = 1$).

On the other hand, pooling layers are not translationally equivariant. Note that pooling layers extract the feature corresponding to large entries from the convolution matrix at the cost of precise spatial information of the feature as in the following example.

Example 10.6.1 ($p(X)$ is not translationally equivariant). Consider a 4×4 matrix:

$$X = \begin{pmatrix} 2 & 2 & 0 & 1 \\ 2 & 2 & 0 & 1 \\ 2 & 2 & 0 & 1 \\ 2 & 2 & 0 & 1 \end{pmatrix}. \tag{10.46}$$

Decomposing X into 2×2 block matrices, after pooling, we obtain a matrix:

$$p(X) = \begin{pmatrix} 2 & 1 \\ 2 & 1 \end{pmatrix}. \tag{10.47}$$

Now consider the horizontal shift of X,

$$T_h(X) = \begin{pmatrix} 1 & 2 & 2 & 0 \\ 1 & 2 & 2 & 0 \\ 1 & 2 & 2 & 0 \\ 1 & 2 & 2 & 0 \end{pmatrix}. \tag{10.48}$$

After pooling, we obtain a different 2×2 matrix:

$$p(T_h(X)) = \begin{pmatrix} 2 & 2 \\ 2 & 2 \end{pmatrix}. \tag{10.49}$$

Since $p(T_h(X)) \neq T_h(p(X))$, the pooling layer $p(X)$ is not translationally equivariant. Note that if (10.49) were

$$\begin{bmatrix} 1 & 2 \\ 1 & 2 \end{bmatrix}, \tag{10.50}$$

this would imply translational equivariance for the particular matrix X.

Example 10.6.2 ($h_i(x)$ is not translationally equivariant). Consider the following matrix A and vector v:

$$A = \begin{pmatrix} 1 & 0 & 0 \\ 0 & 2 & 0 \\ 0 & 0 & 3 \end{pmatrix}, \quad v = \begin{pmatrix} 1 \\ 0 \\ 0 \end{pmatrix}. \tag{10.51}$$

If we consider v as a 3×1 matrix, then we may apply the vertical shift operator T_v. By straightforward calculation,

$$Av = \begin{pmatrix} 1 \\ 0 \\ 0 \end{pmatrix}, \tag{10.52}$$

$$A(T_v(v)) = \begin{pmatrix} 1 & 0 & 0 \\ 0 & 2 & 0 \\ 0 & 0 & 3 \end{pmatrix} \begin{pmatrix} 0 \\ 1 \\ 0 \end{pmatrix} = \begin{pmatrix} 0 \\ 2 \\ 0 \end{pmatrix}. \tag{10.53}$$

Again $T_v(Av) \neq A(T_v(v))$. This shows that the linear function $f_A(v) := Av$ defined on 3×1 matrices is not equivariant in the sense of Definition 10.6.2. To provide a connection with DNNs, observe that since the layer functions $h_i(X)$ (see Definition 4.1.2) are defined by multiplication by a matrix A of parameters, these functions are also not in general translationally equivariant.

Equivariance of each layer is a desirable property because it can lead to the property of invariance of the classifier. Following Definition 10.6.2, we introduce the following definition.

Definition 10.6.3. Let $\mathbb{R}^{N\times M}$ and be a space of $N \times M$ matrices. Let $f : \mathbb{R}^{N\times M} \to \{0,1\}$ be a classifier function. Then the function f is translationally invariant if

$$f \circ T_h(A) = f(A) \quad \text{and} \quad f \circ T_v(A) = f(A) \tag{10.54}$$

for each $A \in \mathbb{R}^{N\times M}$.

Translational invariance is a desirable property in image classification. For example, if a CNN ϕ is tasked with classifying images of cats and dogs (corresponding to classes 0 and 1, respectively), then we would naturally hope that if A is an image corresponding to a dog, $T_h(A)$ is classified in the same way since it obviously corresponds to the same dog. Mathematically we hope that the classifier ϕ is invariant under both T_v and T_h as given by the definition above:

$$\phi(T_h(A)) = \phi(T_v(A)) = \phi(A) = 1. \tag{10.55}$$

In practice, CNNs tend to behave well with respect to translations in many situations, even when they are not strictly speaking translationally invariant. Furthermore, translational invariance cannot be ensured by just the use of some convolutional layers that are translationally equivariant or even almost translationally equivariant.

We present below a simple, if somewhat artificial, example illustrating how one may use the translational equivariance of CNNs to build a network that is actually translationally invariant.

Example 10.6.3. We consider a dataset of 2D grayscale images with $N \times N$ pixels. Our input hence consists of some finite subset of $\mathbb{R}^{N\times N}$. Our goal is to classify the images in the dataset into κ given classes.

We are going to make use of padding layers to change the almost translational invariance of CNNs into an exact translational invariance. We start our DNN by alternating d padding layer functions P_k with d convolutional functions C_k for $k = 1, \ldots, d$, as given by Definition 10.2.1. We hence have d kernels K_k of dimension $D_k \times D_k$, while each padding layer P_k has padding size $2s_k$.

Starting with a matrix of size N, we apply the first padding function to obtain a matrix of size $N+2s_1$. We follow with the convolutional function c_1, yielding a matrix of size $N_1 = N+2s_1-D_1+1$. Each combination of a padding function and a convolutional function hence yields matrices of size N_k with $N_0 = N$ and $N_k = N_{k-1} + 2s_k - D_k + 1$ for $k \geq 1$.

We could consider convolutional layers with various input or output channels, but to keep things as simple as possible in this example we assume that all convolutional layers, *except the last one*, only have one input and output channel. Our last convolutional layer for $k = d$ has one input channel but κ (the number of classes) output channels and is hence represented by κ kernels $K_{d,i}$ for $i = 1, \ldots, \kappa$. We assume for simplicity that all kernels $K_{d,i}$ have the same dimension D_d. See Fig. 10.7 for an illustration of such an ANN.

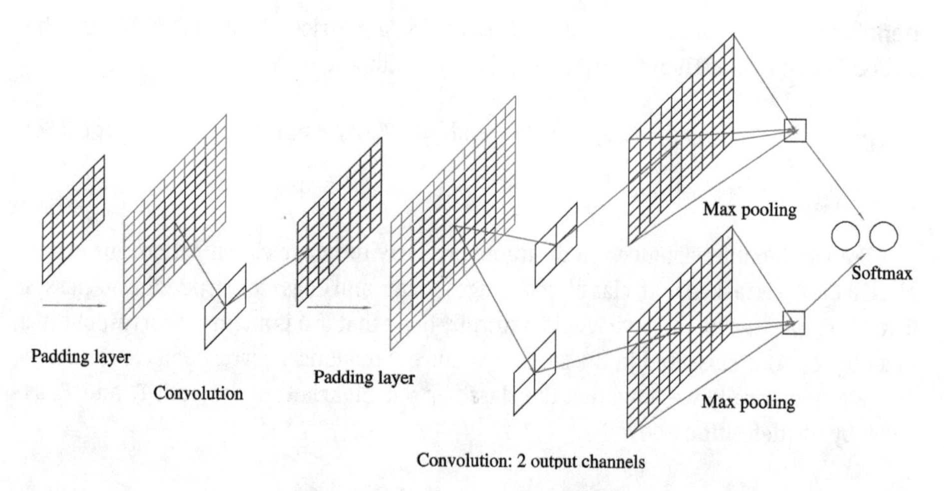

Figure 10.7: A representation of a simple network with translation invariance for two classes.

Applied on a matrix $M_0 \in \mathbb{R}^{N\times N}$, the combination of the layer functions yields κ matrices $M^i(M)$, $i = 1, \ldots, K$, of dimension $N_d \times N_d$ with

$$N_d = N + 2s_1 + \cdots + 2s_d - D_1 - \cdots - D_d + d.$$

We then apply a max pooling layer. In this case, we take the simplest possible such layer by taking the maximum over the whole $N_d \times N_d$ matrix. In mathematical terms, to our κ matrices M^i, this max pooling layer associates κ numbers q_i for $i = 1, \ldots, \kappa$ with

$$q_i = \max_{\alpha,\beta=1,\ldots,N_d} M^i_{\alpha\beta}.$$

This max pooling layer will change the translational equivariance of the previous padding and convolutional layers into translation invariance of this network with additional padding layers.

We finally apply a softmax layer function which transforms the vector $(q_1, \dots, q_K)$ into a normalized probability distribution $(p_1, \dots, p_K) = (p_1(M), \dots, p_K(M))$ for all $M \in \mathbb{R}^{N \times N}$.

Let us now check that the resulting classifier $(p_1(M), \dots, p_K(M))$ is translationally invariant, under some appropriate conditions on all parameters. Consider for instance the horizontal shift operator T_h. Then our goal is to show that $p_i(T_h(M)) = p_i(M)$ for all $i = 1, \dots, K$.

We first assume that the initial images were themselves padded on the right so that $M_{\alpha,\beta} = 0$ for $\beta = N$ and all $1 \le \alpha \le N$. This ensures that $T_h(M)$ contains the same information as M: Any feature present in M is still present in $T_h(M)$. It also ensures that padding the initial image is equivariant through T_h, namely $P_1(T_h(M)) = T_h(P_1(M))$.

We may consequently use the translational equivariance of convolutional functions and deduce that

$$(c_1 \circ P_1 \circ T_h(M))_{\alpha\beta} = (T_h \circ c_1 \circ P_1(M))_{\alpha\beta}, \quad \text{for all } \alpha \text{ and all } \beta \ge 2. \tag{10.56}$$

Moreover, assume that $s_1 \ge D_1$, that is, the padding is larger than the kernel size. Then we also immediately have

$$(c_1 \circ P_1 \circ T_h(M))_{\alpha,1} = 0, \quad \text{for all } \alpha.$$

For the same reason, we have

$$(c_1 \circ P_1(M))_{\alpha,N_1} = 0, \quad \text{for all } \alpha. \tag{10.57}$$

By the definition of T_h, this implies

$$(T_h \circ c_1 \circ P_1(M))_{\alpha,1} = 0, \quad \text{for all } \alpha,$$

and therefore we obtain the strict translational equivariance

$$(c_1 \circ P_1 \circ T_h(M))_{\alpha\beta} = (T_h \circ c_1 \circ P_1(M))_{\alpha\beta}, \quad \text{for all } \alpha,\ \beta.$$

We emphasize again that a padding layer with a large enough padding size $s_1 \ge D_1$ is critical for this property.

We also note that (10.57) shows that the last column of $c_1 \circ P_1(M)$ is composed only of 0s, just as we had assumed on M initially. Therefore we can repeat the argument for the next padding and convolutional layer and obtain

$$(c_2 \circ P_2 \circ c_1 \circ P_1 \circ T_h(M))_{\alpha\beta} = (T_h \circ c_2 \circ P_2 \circ c_1 \circ P_1(M))_{\alpha\beta}, \quad \text{for all } \alpha,\ \beta,$$

provided that $s_2 \ge D_2$.

We can keep on repeating this argument on each combination of padding and convolutional layer until the last such layer with the κ output channels. The strict translational equivariance applies to each output channel and we obtain

$$(M^i \circ T_h(M))_{\alpha\beta} = (T_h \circ M^i(M))_{\alpha\beta}, \quad \text{for all } \alpha,\ \beta,\ i,$$

provided that

$$s_k \geq D_k, \quad \text{for all } k = 1, \dots, \kappa. \tag{10.58}$$

Finally we note that the max pooling layer that we are using is translationally invariant since it applies to the whole matrix: one can readily check from the definition of T_h

$$\max_{\alpha,\beta=1,\dots,N_d} M_{\alpha\beta} = \max_{\alpha,\beta=1,\dots,N_d} (T_h M)_{\alpha\beta},$$

for any matrix M.

We can immediately deduce that $q_i \circ T_h(M) = q_i(M)$ and conclude as desired that

$$p_i \circ T_h(M) = p_i(M).$$

We finally note that however useful translational invariance may be, there are in practice often compelling reasons to deviate from this network construction with a large number of padding layers, special max pooling, and no fully connected layers even if we lose invariance. The constraints (10.58) imposing large padding size are demanding and lead to large layers that may be too costly computationally for example. One may also obtain a better accuracy for the classifier by replacing the max pooling layer by a fully connected layer that can be trained but is not translationally invariant.

It is also important to observe that we only proved the invariance under a single shift T_h. To obtain the same result by the application of n shifts $T_h, T_h \circ \cdots \circ T_h$, we would need to ensure that the initial matrices M have their n last columns filled with 0s, which is also much more demanding.

10.7 Summary of CNNs

To summarize this brief presentation of CNNs, we highlight the following:

1. Convolution: CNNs are obtained by alternating convolutional and other types of layers such as pooling or fully connected layers.
2. Weight sharing: Convolutional layers typically have far fewer weights than fully connected layers, which has several advantages ranging from reduced computation time to less overfitting.
3. Locality: CNNs may find local features more easily. The convolutional layers extract the most important features of images by encoding them into large elements of their output matrices.
4. Translational equivariance: CNNs enjoy some robustness with respect to feature location through the partial translational equivariance of their convolutional layers.

10.8 Exercises

Exercise 1 (Basic definition of convolutions). Consider a matrix M in $\mathbb{R}^{N\times M}$ and a matrix K in $\mathbb{R}^{D\times D}$. Recall the definition of the convolution $K * A$ and the condition on the sizes D, N, and M for the operation to be valid.

Exercise 2 (Some examples of convolutions).

i. Calculate $K * M$ for

$$K = \begin{bmatrix} 1 & -1 \\ -1 & 1 \end{bmatrix}, \quad M = \begin{bmatrix} 2 & 2 & -2 \\ 2 & 2 & -2 \\ -2 & -2 & -2 \end{bmatrix}.$$

ii. Calculate $K * M$ for

$$K = \begin{bmatrix} 2 & -1 & -1 \\ -2 & 1 & 1 \\ 0 & 0 & 0 \end{bmatrix}, \quad M = \begin{bmatrix} 2 & 2 & 2 & -2 & -2 \\ 2 & 2 & 2 & 2 & -2 \\ 2 & 2 & 2 & 2 & 2 \\ 100 & 100 & 100 & 100 & 100 \end{bmatrix}.$$

Exercise 3 (Linearity of convolutions). Consider matrices A, B in $\mathbb{R}^{N\times M}$ and two kernels K, L in $\mathbb{R}^{D\times D}$ with $D \le N$ and $D \le M$.

i. Show that

$$(K + L) * A = K * A + L * A \quad \text{and} \quad K * (A + B) = K * A + K * B.$$

ii. For any $\lambda \in \mathbb{R}$, show that

$$(\lambda K) * A = K * (\lambda A) = \lambda (K * A).$$

Exercise 4 (Commuting convolutions).

i. Consider two continuous functions $f, g : \mathbb{R}^n \to \mathbb{R}$ with compact support. Show that for any $\boldsymbol{x} \in \mathbb{R}^n$,

$$(f * g)(\boldsymbol{x}) = (g * f)(\boldsymbol{x}).$$

ii. Consider a matrix A in $\mathbb{R}^{N\times N}$ and a matrix B in $\mathbb{R}^{D\times D}$. Show that we can calculate both the convolution $A * B$ and the convolution $B * A$ only if $N = D$.

iii. In the case where A, B are matrices in $\mathbb{R}^{N\times N}$, show that $A * B \in \mathbb{R}$ and that $A * B = B * A$.

Exercise 5 (Some special cases of convolutions). We consider a matrix M in $\mathbb{R}^{N\times N}$ with $N \ge 2$.

i. Consider a scalar $\lambda \in \mathbb{R}$ which can be seen as a 1×1 matrix $K \in \mathbb{R}^{1\times1}$ with simply $K_{1,1} = \lambda$. Prove that

$$K * M = \lambda M \quad \text{or} \quad (K * M)_{ij} = \lambda M_{ij} \quad \text{for all } 1 \le i, j \le N.$$

ii. For any kernel $K \in \mathbb{R}^{D\times D}$ with $D \le N$, show that $K * M = M$ if and only if $D = 1$ and $K_{1,1} = 1$.

iii. Consider now the kernel $K \in \mathbb{R}^{2\times2}$ given by

$$K = \begin{bmatrix} 1 & 0 \\ 0 & 0 \end{bmatrix}.$$

Show that $K * M$ is equal to a block of M in the following sense:

$$(K * M)_{ij} = M_{ij}, \quad \text{for all } 1 \le ij \le N - 1.$$

Exercise 6 (The identity matrix and convolutions). Consider a matrix M in $\mathbb{R}^{N\times M}$ and the identity matrix I in $\mathbb{R}^{D\times D}$ for $D > 1$, $D \le N$, and $D \le M$. Explain what is the expression for $I * M$ and show that $I * M$ is always different from M.

Exercise 7 (Padding is linear). Consider matrices $A, B \in \mathbb{R}^{N\times N}$, some scalar $\lambda \in \mathbb{R}$, and a padding layer function P with padding size $2s$. Show that

$$P(A + \lambda B) = P(A) + \lambda P(B).$$

Exercise 8 (Max pooling may lose some information).

i. Consider a matrix $M \in \mathbb{R}^{N\times N}$ where all entries are non-negative: $M_{ij} \ge 0$ for all $1 \le ij \le N$. Assume $N = SD$ and consider further a max pooling layer function $p : \mathbb{R}^{N\times N} \to \mathbb{R}^{S\times S}$. Prove that if $p(M) = 0$, then $M = 0$.

ii. Give an example of matrix $M \in \mathbb{R}^{6\times6}$, with possibly negative entries, such that one has $p(M) = 0$ but $M \neq 0$ for the max pooling layer function $p : \mathbb{R}^{6\times6} \to \mathbb{R}^{3\times3}$.

Exercise 9 (Max pooling is not linear). We consider matrices $A, B \in \mathbb{R}^{N\times N}$ with $N = SD$ and a max pooling layer function $p : \mathbb{R}^{N\times N} \to \mathbb{R}^{S\times S}$.

i. For any $\lambda \ge 0$, show that

$$p(\lambda M) = \lambda p(M).$$

ii. In the special case where $D = N$, show that

$$p(-M) = p(M)$$

only if $\min_{1\le i,j\le N} M_{ij} = -\max_{1\le i,j\le N} M_{ij}$.

iii. Find an example for $N = 4, D = 2$ of matrices A, B such that

$$p(A + B) \neq p(A) + p(B).$$

Exercise 10 (Revisiting translational equivariance). Consider a matrix kernel K of size $s \times s$ which we will apply to grayscale images via convolution.

i. For an image represented by a matrix A of size $N \times N$, explain how to obtain the convolution $B = K * A$ and give its size.
ii. Let D be a matrix of size $M \times M$ with $M < N$. Suppose that $N - M > 0$ is an even number so that $N - M = 2t$ for an integer $t > 0$. Recall how we may transform the matrix D into a larger matrix C of size $N \times N$ by adding 0 in a symmetric manner.
iii. Given a matrix A of size $N \times N$, we define a matrix $\tau_{k_0,\ell_0}A$ by indices k_0, ℓ_0 in the following manner:

$$\tau_{k_0,\ell_0}A_{ij} = A_{i-k_0,j-l_0}, \quad \text{for } \max 0, k_0 \leq i \leq \min(m, m + k_0)$$
$$\text{and } \max 0, \ell_0 \leq j \leq \min(n, n + \ell_0),$$

and $\tau_{k_0,\ell_0}A_{ij} = 0$ in all other cases. Explain why if A was obtained from a matrix D as in the previous question with $|k_0|, |\ell_0| \leq t$, then the image corresponding to $\tau_{k_0,l_0}A_{ij}$ is the translation of the image corresponding to A.
iv. Assume again that the matrix A is obtained through a matrix D from question 2 with now $|k_0|, |l_0| \leq t$. Show that

$$K * (\tau_{k_0,\ell_0}A) = \tau_{k_0,\ell_0}(K * A).$$

Explain why this property is useful when doing image processing.

A Review of the chain rule

Recall that the derivative of a composition of two functions can be calculated using the following theorem.

Theorem A.0.1. *If $f, g : \mathbb{R}$ are differentiable functions, then $f \circ g : \mathbb{R} \to \mathbb{R}$ is also differentiable and*

$$\frac{d}{dx} f \circ g(x) = \frac{df}{dy}(y)\bigg|_{y=g(x)} \frac{dg}{dx}(x) = f'(g(x))g'(x). \tag{A.1}$$

Theorem A.0.1 is called the *chain rule*, and it is central to backpropagation.

Recall from Definition 4.1.3 that DNNs are typically composed not of functions from $\mathbb{R}$ to $\mathbb{R}$, but of functions from $\mathbb{R}^{n_i}$ to $\mathbb{R}^{n_{i+1}}$, where $n_i, n_{i+1} \geq 1$. Therefore, we introduce the multivariable chain rule.

Theorem A.0.2. *If $g : \mathbb{R}^{n_1} \to \mathbb{R}^{n_2}$ and $f : \mathbb{R}^{n_2} \to \mathbb{R}^{n_3}$ are differentiable, then $f \circ g : \mathbb{R}^{n_1} \to \mathbb{R}^{n_3}$ is also differentiable and its partial derivatives are given by*

$$\frac{\partial}{\partial x_i} f \circ g(\boldsymbol{x}) = \frac{\partial f}{\partial y_1}(\boldsymbol{y})\bigg|_{y=g(x)} \frac{\partial g_1}{\partial x_i}(\boldsymbol{x}) + \cdots + \frac{\partial f}{\partial y_{n_2}}(\boldsymbol{y})\bigg|_{y=g(x)} \frac{\partial g_{n_2}}{\partial x_i}(\boldsymbol{x}) \tag{A.2}$$

$$= \sum_{j=1}^{n_2} \frac{\partial f}{\partial y_j}(\boldsymbol{y})\bigg|_{y=g(x)} \frac{\partial g_j}{\partial x_i}(\boldsymbol{x}). \tag{A.3}$$

Observe that for each i, $\frac{\partial}{\partial x_i} f \circ g(\boldsymbol{x})$ is a vector with n_3 components because $f \circ g : \mathbb{R}^{n_1} \to \mathbb{R}^{n_3}$:

$$\frac{\partial}{\partial x_i} f \circ g(\boldsymbol{x}) = \begin{pmatrix} \frac{\partial}{\partial x_i} f_1 \circ g(\boldsymbol{x}) \\ \vdots \\ \frac{\partial}{\partial x_i} f_{n_3} \circ g(\boldsymbol{x}) \end{pmatrix}. \tag{A.4}$$

We can arrange all the partial derivatives of $f \circ g$ into the columns of an $n_3 \times n_1$ matrix:

$$\frac{d}{d\boldsymbol{x}} f \circ g(\boldsymbol{x}) = \begin{pmatrix} \frac{\partial}{\partial x_1} f_1 \circ g(\boldsymbol{x}) & \cdots & \frac{\partial}{\partial x_{n_1}} f_1 \circ g(\boldsymbol{x}) \\ \vdots & & \vdots \\ \frac{\partial}{\partial x_1} f_{n_3} \circ g(\boldsymbol{x}) & \cdots & \frac{\partial}{\partial x_{n_1}} f_{n_3} \circ g(\boldsymbol{x}) \end{pmatrix}. \tag{A.5}$$

From (A.3), we see that the matrix $\frac{d}{d\boldsymbol{x}} f \circ g(\boldsymbol{x})$ can be calculated via the following matrix multiplication:

https://doi.org/10.1515/9783111025551-011

$$\frac{d}{dx}f \circ g(\boldsymbol{x}) = \begin{pmatrix} \frac{\partial}{\partial x_1} f_1 \circ g(\boldsymbol{x}) & \cdots & \frac{\partial}{\partial x_{n_1}} f_1 \circ g(\boldsymbol{x}) \\ \vdots & & \vdots \\ \frac{\partial}{\partial x_1} f_{n_3} \circ g(\boldsymbol{x}) & \cdots & \frac{\partial}{\partial x_{n_1}} f_{n_3} \circ g(\boldsymbol{x}) \end{pmatrix} \tag{A.6}$$

$$= \begin{pmatrix} \frac{\partial f_1}{\partial y_1}(\boldsymbol{y}) & \cdots & \frac{\partial f_1}{\partial y_{n_2}}(\boldsymbol{y}) \\ \vdots & & \vdots \\ \frac{\partial f_{n_3}}{\partial y_1}(\boldsymbol{y}) & \cdots & \frac{\partial f_{n_3}}{\partial y_{n_2}}(\boldsymbol{y}) \end{pmatrix}\Bigg|_{\boldsymbol{y}=g(\boldsymbol{x})} \begin{pmatrix} \frac{\partial g_1}{\partial x_1}(\boldsymbol{x}) & \cdots & \frac{\partial g_1}{\partial x_{n_1}}(\boldsymbol{x}) \\ \vdots & & \vdots \\ \frac{\partial g_{n_2}}{\partial x_1}(\boldsymbol{x}) & \cdots & \frac{\partial g_{n_2}}{\partial x_{n_1}}(\boldsymbol{x}) \end{pmatrix}. \tag{A.7}$$

Observe that the first matrix is $\frac{df}{dy}(\boldsymbol{y})|_{\boldsymbol{y}=g(\boldsymbol{x})}$, while the second is $\frac{dg}{dx}(\boldsymbol{x})$. Therefore,

$$\frac{d}{dx}f \circ g(\boldsymbol{x}) = \frac{df}{dy}(\boldsymbol{y})\bigg|_{\boldsymbol{y}=g(\boldsymbol{x})} \frac{dg}{dx}(\boldsymbol{x}), \tag{A.8}$$

which has the same form as the single variable chain rule (A.1).

Bibliography

[1] J. Alman and V. Williams. A refined laser method and faster matrix multiplication. In *Proceedings of the 2021 ACM-SIAM Symposium on Discrete Algorithms (SODA)*, pages 522–539, 2021.

[2] F. Anselmi, J. Leibo, L. Rosasco, J. Mutch, A. Tacchetti, and T. Poggio. Unsupervised learning of invariant representations. *Theoretical Computer Science*, 633:112–121, 2016.

[3] B. E. Boser, I. M. Guyon, and V. N. Vapnik. A training algorithm for optimal margin classifiers. In *Proceedings of the Fifth Annual Workshop on Computational Learning Theory, COLT '92*, pages 144–152, New York, NY, USA, July 1992. Association for Computing Machinery, 1992.

[4] L. Bottou. Large-scale machine learning with stochastic gradient descent. In *Proceedings of COMP-STAT'2010*, pages 177–186, 2010.

[5] H. Brezis. *Functional Analysis, Sobolev Spaces and Partial Differential Equations*. Springer, 2011.

[6] J. Bruna and S. Mallat. Invariant scattering convolution networks. *IEEE Transactions on Pattern Analysis and Machine Intelligence*, 35:1872–1886, 2013.

[7] J. Bruna, A. Szlam, and Y. LeCun. Learning stable group invariant representations with convolutional networks. In *ICLR 2014*, page 13, 2014.

[8] J. Buhmann and H. Kuhnel. Unsupervised and supervised data clustering with competitive neural networks. In *IJCNN International Joint Conference on Neural Networks*, vol. 4, pages 796–801, 1992.

[9] J. S. Cramer. The origins of logistic regression, 2002.

[10] M. K. David and G. Kleinbaum. *Logistic Regression. Statistics for Biology and Health*. Springer, New York, NY, 3rd edition, 2010.

[11] B. Després. *Neural Networks and Numerical Analysis*. De Gruyter, 2022.

[12] A. Fawzi, M. Balog, A. Huang, T. Hubert, B. Romera-Paredes, M. Barekatain, A. Novikov, F. J. R. Ruiz, J. Schrittwieser, G. Swirszcz, and et al.. Discovering faster matrix multiplication algorithms with reinforcement learning. *Nature*, 610(7930):47–53, 2022.

[13] K. Fukushima. Neocognitron: A self-organizing neural network model for a mechanism of pattern recognition unaffected by shift in position. *Biological Cybernetics*, 36(4):193–202, Apr. 1980.

[14] F. Galton. Kinship and correlation. *The North American Review*, 150(401):419–431, 1890.

[15] M. Gaviano. Some general results on convergence of random search algorithms in minimization problems. In *Towards Global Optimization*, pages 149–157. North Holland, Amsterdam, 1975.

[16] X. Glorot, A. Bordes, and Y. Bengio. Deep sparse rectifier neural networks. In G. Gordon, D. Dunson and M. Dudík, editors, *Proceedings of the Fourteenth International Conference on Artificial Intelligence and Statistics*, Fort Lauderdale, FL, USA, 11–13 Apr. 2011, pages 315–323. Proceedings of Machine Learning Research, vol. 15. PMLR, 2011.

[17] I. Goodfellow, Y. Bengio, and A. Courville. *Deep Learning*. MIT Press, Nov. 2016. Google-Books-ID: Np9SDQAAQBAJ.

[18] G. Hinton and T. Sejnowski. *Unsupervised Learning: Foundations of Neural Computation*. MIT Press, 1999.

[19] K. Hornik. Approximation capabilities of multilayer feedforward networks. *Neural Networks*, 4(2):251–257, Jan. 1991.

[20] D. H. Hubel and T. N. Wiesel. Receptive fields of single neurones in the cat's striate cortex. *The Journal of Physiology*, 148:574–591, Oct. 1959.

[21] D. H. Hubel and T. N. Wiesel. Receptive fields, binocular interaction and functional architecture in the cat's visual cortex. *The Journal of Physiology*, 160:106–154, Jan. 1962.

[22] J. Jumper, R. Evans, A. Pritzel, T. Green, M. Figurnov, O. Ronneberger, K. Tunyasuvunakool, R. Bates, A. Žídek, A. Potapenko, A. Bridgland, C. Meyer, S. A. A. Kohl, A. J. Ballard, A. Cowie, B. Romera-Paredes, S. Nikolov, R. Jain, J. Adler, T. Back, S. Petersen, D. Reiman, E. Clancy, M. Zielinski, M. Steinegger, M. Pacholska, T. Berghammer, S. Bodenstein, D. Silver, O. Vinyals, A. W. Senior, K. Kavukcuoglu, P. Kohli, and D. Hassabis. Highly accurate protein structure prediction with AlphaFold. *Nature*, 596(7873):583–589, Aug. 2021.

https://doi.org/10.1515/9783111025551-012

[23] J. Kiefer and J. Wolfowitz. Stochastic estimation of the maximum of a regression function. *The Annals of Mathematical Statistics*, 23(3):462, 1952.

[24] T. Kohonen. *Self-Organizing Maps*. Springer, Berlin, Heidelberg, 2001.

[25] A. N. Kolmogorov and S. V. Fomin. *Elements of the Theory of Functions and Functional Analysis*. Dover, 1999.

[26] A. Krizhevsky, I. Sutskever, and G. E. Hinton. ImageNet classification with deep convolutional neural networks. *Communications of the ACM*, 60(6):84–90, May 2017.

[27] Y. LeCun, Y. Bengio, and G. Hinton. Deep learning. *Nature*, 521:436–444, 2015.

[28] Y. LeCun, B. Boser, J. S. Denker, D. Henderson, R. E. Howard, W. Hubbard, and L. D. Jackel. Backpropagation applied to handwritten zip code recognition. *Neural Computation*, 1(4):541–551, Dec. 1989.

[29] Y. LeCun, B. Boser, J. S. Denker, D. Henderson, R. E. Howard, W. Hubbard, and L. D. Jackel. Backpropagation applied to handwritten zip code recognition. *Neural Computation*, 1(4):541–551, 1989.

[30] Y. LeCun, L. Bottou, G. Orr, and K. Müller. *Efficient BackProp*. Springer, Berlin Heidelberg, 2012.

[31] S. Mallat. Group invariant scattering. *Communications on Pure and Applied Mathematics*, 65(10):1331–1398, 2012.

[32] S. Mallat. Understanding deep convolution networks. *Philosophical Transactions of the Royal Society A*, 374:20150203, 2016.

[33] D. Marr and T. Poggio. A computational theory of human stereo vision. *Proceedings of the Royal Society of London. Series B, Biological Sciences*, 204(1156):301–328, 1979.

[34] W. McCulloch and W. Pitts. A logical calculus of the ideas immanent in nervous activity. *Bulletin of Mathematical Biophysics*, 5:115–133, 1943.

[35] M. Minsky and S. Papert. *Perceptrons: An Introduction to Computational Geometry*. MIT Press, Cambridge, 1988.

[36] Y. Nesterov. *Introductory Lectures on Convex Optimization: A Basic Course*. Springer, New York, NY, 2004.

[37] P. B. Palmer and D. G. O'Connell. Regression analysis for prediction: Understanding the process. *Cardiopulmonary Physical Therapy Journal*, 20(3):23, 2009.

[38] A. Pinkus. Approximation theory of the MLP model in neural networks. *Acta Numerica*, 8:143–195, 1999. arXiv:1910.13029.

[39] A. Pinkus. Weierstrass and approximation theory. *Journal of Approximation Theory*, 107(1):1–66, Nov. 2000.

[40] B. Polyak. *Introduction to Optimization*. Optimization Software, Inc., New York, 1987.

[41] N. Qian. On the momentum term in gradient descent learning algorithms. *Neural Networks*, 12(1):145–151, 1999.

[42] H. Robbins and S. Monro. A stochastic approximation method. *The Annals of Mathematical Statistics*, 22(3):400, 1951.

[43] F. Rosenblatt. The perceptron: A probabilistic model for information storage and organization in the brain. *Psychological Review*, 10(9):386–408, 1958.

[44] W. Rudin. *Functional Analysis*. McGraw-Hill, 1973.

[45] B. Schölkopf, A. J. Smola, F. Bach, and et al.. *Learning with Kernels: Support Vector Machines, Regularization, Optimization, and Beyond*. MIT Press, 2002.

[46] T. Serre, L. Wolf, and T. Poggio. Object recognition with features inspired by visual cortex. In *2005 IEEE Computer Society Conference on Computer Vision and Pattern Recognition (CVPR'05)*, pages 994–1000, vol. 2, June 2005. 1063-6919.

[47] J. Spall. In *Introduction to Stochastic Search and Optimization: Estimation, Simulation, and Control*. Wiley, NJ, 2003.

[48] J. F. Steffensen. *Interpolation*. Dover Books on Mathematics. Dover Publications, Mineola, NY, 2nd edition, Mar. 2006.

[49] G. Strang. The functions of deep learning. *SIAM News*, 51(10), 2018.

[50] V. Strassen. Gaussian elimination is not optimal. *Numerische Mathematik*, 13(4):354–356, 1969.

[51] C. Szegedy, W. Liu, Y. Jia, P. Sermanet, S. Reed, D. Anguelov, D. Erhan, V. Vanhoucke, and A. Rabinovich. Going deeper with convolutions, Sept. 2014. arXiv:1409.4842 [cs].

[52] V. Vapnik. Pattern recognition using generalized portrait method. *Automation and Remote Control*, 24:774–780, 1963.

[53] V. Vapnik and A. Chervonenkis. *Theory of Pattern Recognition*, 1974.

[54] G. U. Yule. On the theory of correlation. *Journal of the Royal Statistical Society*, 60(4):812–854, 1897.

Index

https://doi.org/10.1515/9783111025551-013

Printed in the USA
CPSIA information can be obtained
at www.ICGtesting.com
LVHW081526121223
766314LV00002B/25

9 783111 024318